PENGUIN BOOKS

THE PENGUIN BOOK OF GARDEN WRITING

David Wheeler began gardening when he was four, and has continued to do so ever since except for a brief career on the high seas when he was a teenager. He is the author of *Over the Hills from Broadway: Images of Cotswold Gardens* (1991) and *Panoramas of English Gardens* (1991). He founded the literary gardening journal *Hortus* in 1987, which he continues to edit from his home near Presteigne on the England–Wales border. He has previously edited two anthologies of garden writing from the early volumes of *Hortus*: *By Pen and by Spade* (1990) and *The Generous Garden* (1991). He has lectured on gardens and garden-making in England, America and Canada. He writes regularly for *Country Life* magazine on ornamental gardens, the cultivation of fruit and vegetables, and the British countryside. He contributes on a freelance basis to a range of other publications, including *The Garden* (the journal of the Royal Horticultural Society), *Horticulture* magazine in the USA, and several British newspapers and periodicals.

The illustrator, Simon Dorrell, is a painter who regularly exhibits his work in London and other British towns. Exhibitions of his paintings have also been seen in New York and Zurich. He is also a book illustrator and Art Editor of *Hortus*. The pen and ink drawings in this book are based on scenes at his and David Wheeler's garden, Bryan's Ground, in north-west Herefordshire, which is now open to the public.

In St. Ives,
Bryan's Ground.
Simon Dorrell.

THE PENGUIN BOOK OF

GARDEN WRITING

EDITED BY

David Wheeler

PENGUIN BOOKS

PENGUIN BOOKS

Published by the Penguin Group
Penguin Books Ltd, 27 Wrights Lane, London W8 5TZ, England
Penguin Putnam Inc., 375 Hudson Street, New York, New York 10014, USA
Penguin Books Australia Ltd, Ringwood, Victoria, Australia
Penguin Books Canada Ltd, 10 Alcorn Avenue, Toronto, Ontario, Canada M4V 3B2
Penguin Books (NZ) Ltd, 182–190 Wairau Road, Auckland 10, New Zealand

Penguin Books Ltd, Registered Offices: Harmondsworth, Middlesex, England

This collection first published by Viking 1996
Published in Penguin Books 1998
6

Introduction, selection and editorial matter copyright © David Wheeler, 1996
Illustrations copyright © Simon Dorrell, 1996
All rights reserved

The moral right of the editor has been asserted

Printed in England by Clays Ltd, St Ives plc

Dedicated with much
appreciation and affection to
Simon Dorrell

CONTENTS

ACKNOWLEDGEMENTS

My thanks are due to Mary Bland, Angela Boschi, Anne Carter, Liz Robinson, Katherine Swift and Robin Whalley, friends who have shown a lively interest in the compilation of this book, and who suggested authors and books I might not otherwise have discovered. I should also like to thank the staff of the Royal Horticultural Society's Lindley Library, and Alan Martin and the librarian of the Imperial War Museum for their rapid response to my varied appeals. I should also like to express my appreciation to Simon Dorrell for his practical help at every stage, and to Rosemary Verey, who gave me free range over her library shelves, saving me many a long trip to London.

I also owe a debt of thanks to the following people and organizations for allowing me to reprint material which remains in copyright:

John Akeroyd, Jane Allsopp, Ronald Blythe, Beth Chatto, Thomas Christopher, G. F. Dutton, Joe Eck and Wayne Winterrowd, Robin Lane Fox, John Francis, Nancy-Mary Goodall, Jim Gould, Patti Hagan and *The Wall Street Journal*, Deborah Kellaway, John Kelly, Stephen Lacey, Dawn Macleod, Virginia Mitchell, Mirabel Osler, Peter Parker, Timothy Ribbesford, Liz Robinson, Josephine Swinscow, Graham Stuart Thomas (for his own and on behalf of A. T. Johnson), Derek Toms, Rosemary Verey.

Ethel Armitage: to Reed Consumer Books for extracts from *A Country Garden* (Country Life, 1936)

H. E. Bates: to Laurence Pollinger Ltd on behalf of the Estate of H. E. Bates for an extract from *A Love of Flowers* (Michael Joseph, 1971)

Reginald Blomfield: to B. T. Batsford Ltd for extracts from *The Formal Garden in England* (1892)

Karel Capek: to George Allen & Unwin, an imprint of HarperCollins Publishers Ltd for extracts from *The Gardener's Year* (1931)

Humphrey John Denham: to Chambers Harrap Publishers Ltd for extracts from *The Skeptical Gardener* (Harrap, 1940)

Mrs C. W. Earle: to John Murray (Publishers) Ltd for extracts from *Pot-Pourri from a Surrey Garden* (1897)

Frederick Eden: to Reed Consumer Books for extracts from *A Garden in Venice* (Newnes, 1903)

Margery Fish: to Faber & Faber Ltd for extracts from *Cottage Garden Flowers* (1961) and *A Flower for Every Day* (1965), to Reed Consumer Books Ltd for extracts from *We Made a Garden* (Mitchell Beazley, 1995; first published by Faber, 1956). Text © Lesley Ann Boyd-Carpenter

H. L. V. Fletcher: to Laurence Pollinger Ltd for extracts from *Purest Pleasure* (Hodder, 1948)

Geoffrey Grigson: to David Higham Associates Ltd for an extract from *Gardenage: or The Plants of Ninhursaga* (Routledge, 1952)

Alice Harding: to Sagapress/Timber Press for an extract from *The Peony* (1993)

Thomas Hardy: to Macmillan General Books for extracts from *The Trumpet-Major* (1880) and *Tess of the d'Urbervilles* (1891)

Jason Hill: to Mrs Jillian Leech for extracts from *The Curious Gardener* (Faber, 1932) and *The Contemplative Gardener* (Faber, 1940)

William Bowyer Honey: to Faber & Faber Ltd for an extract from *Gardening Heresies and Devotions* (1939)

C. S. Jarvis: to Reed Consumer Books for an extract from *Gardener's Medley* (Country Life, 1951)

Gertrude Jekyll: to Addison Wesley Longman Ltd for extracts from *Wood and Garden* (1899) and *Home and Garden* (1900)

Rudyard Kipling: to A. P. Watt Ltd on behalf of the National Trust for Places of Historic Interest or Natural Beauty for 'The Glory of the Garden' from *Rudyard Kipling's Verse: Definitive Edition*

Elizabeth Lawrence: to University of North Carolina Press for extracts from *Through the Garden Gate*, edited by Bill Neal. Copyright © 1990 by the University of North Carolina Press

Clare Leighton: to David Leighton for extracts from *Four Hedges* (Gollancz, 1935)

John Lodwick: to Reed Consumer Books for an extract from *The Asparagus Trench* (William Heinemann, 1960)

Neil McEacharn: to Reed Consumer Books for an extract from *The Villa Taranto* (Country Life, 1954)

A. A. Milne: to Curtis Brown Ltd for an extract from *If I May* (Methuen, 1920). Copyright A. A. Milne, 1920

Beverly Nichols: to Eric Glass Ltd for an extract from *Down the Garden Path* (1932)

Harold Nicolson: to Nigel Nicolson for extracts from *Harold Nicolson's Diaries* (1966–8)

Iris Origo: to John Murray (Publishers) Ltd for extracts from *Images and Shadows* (1970)

Russell Page: to the Executors of the Estate of Russell Page for extracts from *The Education of a Gardener* (Collins, 1962; reprinted by The Harvill Press, 1994) Copyright © the Executors of the Estate of Russell Page, 1994

Joseph Pemberton: to Addison Wesley Longman Ltd for an extract from *Roses: Their History, Development and Cultivation* (1908)

Ruth Pitter: to Mark Pitter for 'The Diehards' (1941)

William Robinson: to John Murray (Publishers) Ltd for extracts from *The Wild Garden* (1870)

Eleanour Sinclair Rohde: to The Medici Society Ltd for an extract from *Herbs and Herb Gardening* (The Medici Society, 1936)

Vita Sackville-West: to Pavilion Books for extracts from *Some Flowers* (1937)

George Sitwell: to David Higham Associates Ltd for extracts from *An Essay on the Making of Gardens* (Murray, 1909)

Norman Thelwell: to Reed Consumer Books for an extract from *A Plank Bridge by a Pool* (Eyre Methuen, 1978)

Arthur Waley: to HarperCollins Publishers Ltd for 'Planning Flowers on the Eastern Embankment', translated by Arthur Waley from *Chinese Poems*

Frank Kingdon Ward: to Random House UK Ltd for an extract from *The Romance of Gardening* (Cape, 1935)

Katharine S. White: to Farrar, Straus & Giroux Inc. for an extract from *Onward and Upward in the Garden*. Copyright © E. B. White, 1979 as Executor of the Estate of Katharine S. White

Louise Beebe Wilder: to Grove/Atlantic Inc. for an extract from *Colour in My Garden* (Atlantic Monthly Press, 1918)

Andrew Young: to the Andrew Young Estate for 'Cornish Flower-Farm'
    from *The Poetical Works of Andrew Young*, edited by Lowbury and Young
    (Secker & Warburg, 1985)

INTRODUCTION

What wond'rous Life in this I lead!
Ripe Apples drop about my head;
The Luscious Clusters of the Vine
Upon my Mouth do crush their Wine;
The Nectaren, and curious Peach,
Into my hands themselves do reach;
Stumbling on Melons, as I pass,
Insnar'd with Flow'rs, I fall on Grass.

From 'The Garden' (c. 1651)
by ANDREW MARVELL

The gardener's footprints wander back through history to earliest times. Whether for food – almost certainly his first need – or for 'physick' to cure his ills, or, later still, for pleasure, he has chipped away at the earth's crust and hacked his way through impenetrable undergrowths to improve conditions for his chosen plants. 'It is wonderful to think,' wrote the best-selling twentieth-century novelist and short-story writer H. E. Bates, during the fear-filled days of the Second World War, 'that one of the few unbroken links between the civilization of ancient Egypt and the civilization of today is the garden.'[1]

For centuries, throughout the world, wherever humans have settled, the ground has been scraped and turned, watered and manured, and made to bear good things. And from a late-twentieth-century

---

1 *A Love of Flowers*, 1971.

viewpoint the vista back through time brims with endeavour, skill and gladful serendipity: amidst the toil and sweat there are sketches of disappointment, failure, frustration and even hatred, but these are always overridden by moments of elation, joy and unrivalled satisfaction, and by glorious friendships. Fortunately for us, much of what has happened has been recorded for posterity by that most useful of gardener's tools, the pen. And it is gardeners' casual accounts, diaries and letters, together with works of horticultural and general literature (including a sprinkling of fiction, poetry and theatrical drama), that have been plundered to fill the pages of this book.

In the early 1930s Eleanour Sinclair Rohde, a prolific writer on plants and gardens in the first half of the twentieth century, remarked upon the curious fact 'that all the knowledge we have of our most ancient gardens is of pleasaunces of great splendour',[2] yet the words and pictures of illuminated manuscripts, and before them of biblical and other religious documents in several languages, point also to a lively peasant husbandry, where the working of the land was essential for survival. This present book, however, aims rather to explore the more decorative aspects of horticulture: those practices which might be separated from others more purely utilitarian by that single most potent of words (to those under its spell) – 'gardening'. And, with one or two exceptions (for it is a huge and different subject), this collection swerves away from 'landscaping', that mightier, more lordly, scale of operation – the 'improvement of grounds', labelled the Picturesque – associated with 'Capability' Brown and his eighteenth-century contemporaries and successors (the most renowned of whom were Richard Payne Knight, Sir Uvedale Price and Humphry Repton) and much commented upon in prose and verse of both a satirical and a laudatory tone. Others, notably John Dixon Hunt and Peter Willis in *The Genius of the Place* (1975), have made their own literary surveys of this movement.

Gardening, although it of course involves the cultivation of fruits and vegetables, on a domestic scale, manages also to reflect the some-

2 *The Story of the Garden*, 1932.

times extravagant (yet always understandable) efforts made by men and
women of all ages to make beautiful the pieces of land they wake up
to every day, whether these be a simple, shared cottage plot where
a few unnamed flowers face the sun, or huge landed estates (referred
to above) where avenues of forest trees ride out from some daunt-
ing architectural façade. And, as Charles Dickens pointed out, the
conditions under which we strive for our effects are common to us
all:

In the culture of flowers there cannot, by their very nature, be anything
solitary or exclusive. The wind that blows over the cottage porch sweeps over
the grounds of the nobleman, and as the rain descends on the just and on the
unjust, so it communicates to all gardens, both rich and poor . . .

Gardening has been a part of my own life for as long as I can remember. I was born at home, near Stroud in Gloucestershire (within shout-
ing distance of the site where, in 1830, the first lawn-mower was
made), on an October morning when there was a heavy frost which,
according to my mother, finished off the beans for that year. My
father, as I have written elsewhere,³ was 'a passionate gardener, bringing form and army order to an undulating hillside'. As a toddler, I
would crawl across the sloping lawn, attracted to bright flowers in
what seemed like neat, far-off borders, and I can remember my first
eye-to-eye contact with a large green caterpillar, at the age of four.
Later, during school years, like many children before and since, I had
my own little patch of ground to do with as I pleased, although my
efforts were not confined to it. As a teenager – after a short spell in the
Merchant Navy when I *couldn't* garden (although I did make forays into
botanic gardens in various ports of call) – and following my father's
early death in 1964, I somehow found the time amid the usual dis-
tractions of youth to keep up the small family garden and make
collections of what, a century earlier, would have been known as
florists' flowers.

³ *Over the Hills from Broadway: Images of Cotswold Gardens*, 1991.

During six years of working in London my garden was one north-facing, three-foot-long window-box (fortuitously, only two miles from the Royal Botanic Gardens, Kew), and it was not until the late 1970s that I acquired my own first 'proper' garden.

This one-third of an acre faced south on free-draining sandy loam on the outskirts of a small market town where Surrey meets Hampshire. This, I learned, was Jekyll country (albeit a dozen miles or so from her beloved Munstead Wood), well-furnished heathland with gorse and pine, cold in winter, almost Mediterranean during a prize summer. Here irises flourished while roses merely languished; wiry, aromatic herbs spread rampantly as thirsty willows and deep-rooted moisture-loving ligularias perished for want of constant water. It was a learning-ground, a place where an untutored yet eagerly enthusiastic gardener, still with other interests, could find his own way. So strong a hold did the gardening 'bug' have on me in those days that when a bad turn of fortune put paid to my job, and left me without an income, I piled my gardening tools into the boot of my small car and went to work in other people's gardens.

Still I remained horticulturally unslaked: I hankered for another dimension. Where, I asked myself, was the contemporary literature of gardening? Who was publishing the vital gardening trivia of our time? Where were to be found the stodge-free musings of lively-minded gardeners, and modern comment on the horticultural quirks of the past? The magazines and journals of the day all seemed to be stuffed with advice, advice and ever more dry advice, and the 'new wave' of garden publishing directed on to the booksellers' shelves a tonnage of glossy picture-books of only superficial interest. So, in 1987, I started my own little quarterly, *Hortus*, which in the ten years of its existence has so far published something like 5,000 pages of original garden writing, from experts and amateurs alike, some of them reprinted here. Thus, garden words became the instruments of my new venture.

As a publisher *and* a gardener I wanted more space, different surroundings; like the seventeenth-century Abraham Cowley (although I could not have expressed it so mellifluously), 'I never had any other desire so strong, and so like to covetousness, as that . . . I might be master at last of a small house and large garden . . ."[4] The long search ended at the gates of a muddy farm track in the middle of Wales. I had found my Paradise: four acres, formal (though neglected) and informal gardens, topiary, dry stone walls retaining generous terraces, a river bank and a stream. Not long after I had installed myself in this new and challenging world, I was joined there by the painter Simon Dorrell.

As I was by no means the first gardener to discover, however, there is likely to be a snake in every Eden. After six years a landowning farmer beckoned to the developers, and our remote and beautiful old house, on a ninth-century monastic site, with a newly reinvigorated garden, was thrown without mercy into an incongruous and uncomfortable new world, and made to join hands with light industry and modern housing.

Fortunately, another house and garden were waiting in the wings. On a November day in 1993 we tore from the frozen Welsh ground as many of our treasured plants as the icy earth would allow, heaved old pieces of stone and garden ornament into the back of removal lorries with our other possessions, tranquillized the dogs and cats, and drove east out of Wales to Bryan's Ground, one mile into England.

Arriving at dusk, we snatched a fragment of time to draw a few tentative lines on the wet snow covering the grass tennis-court, delineating new beds and borders. Like our previous house in Wales, Bryan's Ground had yew hedges, formal and informal spaces. But here also were giant topiary shapes, lawns, a cavernous greenhouse with dilapidated cold frames, the original Edwardian sunk garden with dumpy puddings of clipped box, a folly (a two-storey building which in former times was used to make gas to light the house), and a half-acre kitchen garden walled on two sides with red brick. Here, where someone had gardened happily before, we would carry on.

Our work on the garden at Bryan's Ground (the name recalls a Marcher lord who roamed this part of north-west Herefordshire in

4 Preface to 'The Garden', 1667.

the twelfth century) has led us back in time. We have embraced a style which favours symmetry, small varying enclosures, vistas, an abundance of flower and leaf. Like Elizabethans, we planted an orchard in a linear grid and strewed the grass around the trees with small star-like flowers; we have laid out blocks of one variety of plant (four twelve-foot squares of blue *Iris sibirica* in one scheme), and 'themed' some of the flower-beds according to the rules of colour. But I believe no one can successfully work entirely in the past. Our 'Tudor' herb garden certainly re-creates elements from that creative and distinguished period, but we call it Standen, in honour of Philip Webb's late-nineteenth-century Arts and Crafts house in Sussex, from whose carved overmantles we took a simple design. Our orchard, though formal in its layout, comprises thirty different apples (culinary, cider and dessert) and includes a few modern varieties. Some of our roses were unknown to gardeners fifty, even twenty, years ago, and throughout the garden we have used an eclectic choice of ideas and plants, borrowing all the time from the past – inspired by some of the ideas and descriptions included in this book – but, wherever possible, with a new twist or interpretation. If we have avoided the thoroughly modern or avant-garde, it is because we can remember how silly flared trousers and back-combed hair-dos looked the moment the fashion moved on, and in our fast-changing world a garden (*my* garden), above all else, I think, must conjure a mood of timelessness.

This anthology of garden writing is therefore a personal selection, with perhaps a few prejudices and aspirations firmly underlined. It is, I hope, an entertaining selection which goes some way to reflect the boundless strands of human creativity. If occasionally a dictatorial or arrogant tone is detected it is because passions have been aroused and a strong opinion has been delivered with a sharp-edged pen: such manner of expression today might be considered rude if not tempered with, at least, a fair measure of wit. Too much dogma has crept back into gardening of late, and there is a risk of the pleasure turning sour. If you cannot have fun and express yourself freely within the boundaries of your own hedges and walls, where can you? Gardening is surely the one opportunity open to anyone with a few square yards of ground to do things how they like. There will be – as there always have been – those who disagree with our judgement or 'taste', and would persuade us to do things differently, but a deaf ear can easily be turned to any such discordant voices.

While in a book of modest dimensions it has only been possible to concentrate, with a few exceptions, on gardening literature set down in English, this has by no means restricted it geographically to the British Isles. There are snippets from explorers and plant-hunters in other countries, and from travellers and pioneer emigrants. Nor are the extracts restricted to the words inscribed by 'men of letters'; we hear from patrons *and* workforce, from professionals *and* gifted amateurs, from men and women triumphing over the loss of personal freedom (to contradict John Weathers, who in *My Garden Book* of 1924 affirmed that 'nothing suffers so readily as Gardening when nations are not at peace with one another'); and we hear from that most interesting of human beings, the 'ordinary person' who – out of necessity or a whim to chase dreams – has found it impossible not 'to garden'.

'You cannot treat gardening in a regulation manner,' wrote John D. Sedding towards the close of the nineteenth century. 'It is a discursive subject that of itself breeds laggard humours, inclines you to reverie, and suggests a discursive style.'[5] So a garden begins with an idea, not a plant. It begins, perhaps, with a vision of how a plant might look, whether in isolation (like a forest yew tamed by lowly shears to resemble an ornate bird crouching on the lawn) or massed together with more of its kind in everyone's notion of a colourful and fragrant flower-filled plot. Between the two extremes lie countless variations. While one man's vision is of beans and lettuces, so another's is of pale roses suffocating in their own overpowering perfume. Some gardeners strive for freakish flowers in 'new' colours; others seek variegation of leaf, or plants all in shades of green. One man will want a bower of deep shade to provide a cool refuge on hot afternoons, while his neighbour will heap up stones and rocks and fleck them with cushions

5 *Garden-craft Old and New*, 1890.

of dense foliage to mimic an alpine range. Some gardeners shun anything in yellow, or scented; others want bright orange dahlias with
twelve-inch heads, or leeks the size of vaulting-poles. The garden
writers have seen it all. Fortunes have been made and spent in the
pursuit of gardening, and whole lives have been devoted (and, sometimes, lost) to the cause.

There is a theory, long and strongly held by the English in particular, that English gardens are unrivalled. True, we do have some of the
finest examples ever made (it is gratifying to note that – if visitor
numbers to Hidcote Manor and Sissinghurst Castle are anything to go
by – after some three or four hundred years of garden-making we can
still work the old magic); but what precisely is an 'English garden'?
'Four things are necessary to be provided for,' said Sir William Temple
in 1692: 'Flowers, Fruit, Shade and Water.'[6]

For several centuries many of the plants that flourish under our
grey, maritime skies have been found originally in other lands, and a
stroll around any 'English' garden today will bring an encounter with,
for example, roses from Asia, maples and cherries from Japan, chrysanthemums from China, magnolias from the United States, dahlias
and salvias from Mexico, eucalyptus from Australia, and rosemary and
other tough aromatics from around the Mediterranean. The reason for
this lies in our island history, during which crusaders, soldiers, religious
missionaries, explorers, travellers and men of commerce wandered the
globe, taking a plant from one place and an idea from another, bringing them home, where a complex cross-fertilization took place between the elements of this exotic cargo. So enduring (and, presumably,
desirable) is the idea of an English garden that only ten years ago – at a
time when gardeners in North America were loudly trumpeting the
existence of their own national style – one of America's best-known
botanic gardens, in Chicago, hired the services of an English garden
designer to make an 'English garden'.

Maybe this has happened in other countries, too, although it is hard
to believe, as some books insist, that there is anything more potent for

6 *Upon the Gardens of Epicurus, or Of Gardening, in the Year 1685*, 1692.

a Japanese gardener than a courtyard of smoothly raked gravel, two perfectly-placed stones, a clipped azalea bush and a magnificent flowering cherry. Not that you have to go to the other side of the world to appreciate contrasting styles; here, in what sounds like an unpatriotic moment, is the eighteenth-century French 'Encyclo-paedist' Denis Diderot expounding with enthusiasm on the character-istics of an *English* formal garden, to the disparagement of those of his own country:

We [the French] bring to bear upon the most beautiful situations a ridiculous and paltry taste. The long straight alleys appear to us insipid; the palisades cold and formless. We delight in devising twisted alleys, scroll-work parterres, and shrubs formed into tufts ... It is not so with a neighbouring nation, amongst whom gardens in good taste are as common as magnificent palaces are rare. In England, these kinds of walks, practicable in all weathers, seem made to be the sanctuary of a sweet and placid pleasure; the body is there relaxed, the mind diverted, the eyes are enchanted by the verdure of the turf ... the variety of flowers offers pleasant flattery to the smell and sight, Nature alone, modestly arrayed, and never made up, there spreads out her ornaments and benefits. How the fountains beget the shrubs and beautify them! How the shadows of the woods put the streams to sleep in beds of herbage![7]

'Our England is a garden,' wrote Rudyard Kipling,[8] distilling into rather less than a half-pint pot many a long-winded explication re-garding this country and its people; and to many gardeners through-out the world the splendour of an English garden on a perfect summer's day when, as the poet Matthew Arnold observed, 'musk car-nations break and swell'[9] represents the ultimate horticultural idyll.

This is not the place for a study of the history of English gardening, but I think it would be worth while to set in time and place some of the principal characters responsible for its articulation. Since the

7 *Encyclopédie*, 1751–76.
8 'The Glory of the Garden', 1911.
9 'Thyrsis', 1867.

fourteenth century and the time of Chaucer (whose translation of *The Romaunt of the Rose*, made about 1370, mentions the pomegranate, 'a fruyt wel to lyke') and 'Anon' – who was scribbling paragraphs and stanzas long before then, and who is still putting pen to paper – there has been a long and distinguished procession of epic writers on the garden. Shakespeare created many garden images, and gardeners are among the dramatis personae in some of his plays. Shakespeare's near contemporary Sir Francis Bacon (1561–1626) wrote his celebrated essay 'Of Gardens' (grandly beginning, 'God *Almightie* first Planted a *Garden*') in 1625, forty-five years after Thomas Tusser had brought out his *Five Hundred Pointes of Good Husbandrie*, the first such book to address itself to women. John Gerard's *Herball*, published in 1597, was among the forerunners of many books dealing with the medicinal and culinary properties of plants. Some thirty years later, in 1629, came John Parkinson's massive, punning-titled *Paradisi in Sole Paradisus Terrestris*, dedicated to Henrietta Maria, wife of Charles I, and proclaiming the plants for 'a garden of delight and pleasure'; then, in 1653, Nicholas Culpeper's herbal, rarely out of print in the ensuing three and half centuries.

John Tradescant the Elder (*c.*1570–1638) was gardener to Charles I (among others) and with his son, also John (1608–62), travelled extensively and brought back to England such plants from North America as Virginia creeper and the stag's horn sumach. John the Younger, journeying in North America after his father's death, continued to discover plants suitable for the English climate, including the stately tulip tree, *Liriodendron tulipifera*. The Tradescants lie buried close to the body of Captain Bligh (against whom the crew of the *Bounty*, on a voyage of botanical import, famously mutinied) in the churchyard of St Mary-at-Lambeth, on the south bank of the Thames in central London.

William Lawson, whose *New Orchard and Garden* was published in 1618, marshalled his thoughts with poetic effectiveness, and a rare economy of words; here, with perhaps the poor in mind, he might also be writing a persuasive advertisement for today's Vegetarian Society:

A morsell of bread eaten with comfort, is pleasurable by much than a fat ox with unquietness.

John Evelyn (1620–1706) was eminent both within and beyond gardening circles, leaving diaries which help illuminate an age of immense curiosity and accomplishment. Evelyn's works about trees and the cultivation of salad greens, and a 1693 translation from the French (*The Compleat Gard'ner*), remain cornerstones of his extraordinary output. Alexander Pope, who agitated for the 'amiable simplicity of unadorned nature',[10] was only in his teens when Evelyn died, and was neither the first nor the last to decry the formal layout of gardens. Stephen Switzer, born six years before Pope, favoured straight lines and ordered symmetry, stressing the need for 'boldest Strokes'[11] to connect house and grounds. Until late in his life, when he began business as a seed merchant, Switzer wrote a series of profound books on gardening.

Lancelot Brown (1716–83) earned the nickname 'Capability' from his comments on the 'capabilities' of the estates he was asked to remodel. He is perhaps chiefly remembered for sweeping 'the lawn straight up to the walls'[12] at houses such as Petworth and Blenheim, and for his work at Stowe in Buckinghamshire ('the largest, grandest and most important landscape garden in England'),[13] where he carried out some of William Kent's designs for Lord Cobham.

Humphry Repton, whose life overlapped Brown's by thirty-one years, continued largely in his footsteps, but worked on a smaller scale; he preferred a more closely wooded effect to his landscapes, and had a fondness for rustic, rather than classical, garden buildings. Repton appears under his own name in *Mansfield Park* (1814), where Jane Austen tells us, 'His terms are five guineas a day.' Repton's famed Red Books, the manuscripts and drawings bound in red leather in which he

10 *The Oxford Companion to Gardens*, 1986.
11 ibid.
12 John Dixon Hunt and Peter Willis, *The Genius of the Place*, 1975.
13 John Martin Robinson, *Temples of Delight*, 1990.

presented his schemes to prospective clients, are thought to have numbered more than 400, although many have since been lost. Through Repton we meet the architect John Nash, and are introduced to the Regency period; Nash's enduring contribution to the English skyline (produced after Repton's own designs had been turned down) is the Royal Pavilion at Brighton for the Prince Regent.

Back in the garden proper, so to speak, we find William Cobbett, MP for Oldham, an energetic and highly opinionated writer and social reformer of titanic persuasiveness. His *Rural Rides* of 1830 depicts the English countryside as seen without rose-tinted spectacles, and gives tantalizing glimpses over garden walls. His instantly popular book *The English Gardener* (an adaptation of an earlier work written during a period of self-exile in America, published in 1829) contains pages and pages of shrewd and straightforward advice on every conceivable aspect of practical gardening, including vegetables, vines, nuts and fruit, flowers and trees.

Next we meet a Scotsman who was described by an American landscape gardener of the mid nineteenth century, A. J. Downing, as 'the most distinguished gardening author of the age': John Claudius Loudon. Loudon not only wrote many important gardening books, but compiled the *Encyclopaedia of Gardening* (published in 1822, with many subsequent editions) and in 1826 founded the *Gardener's Magazine*, a pioneering periodical which ran for seventeen years, until his death in 1843. When Loudon married Jane Webb (twenty-four years his junior) in 1830, they became an influential partnership, setting up home in London's Bayswater district, where Jane (who had earlier written a futuristic fantasy entitled *The Mummy*) not only helped her husband with his formidable output of articles and books, but herself wrote almost twenty volumes on gardening, specifically addressing female readers.

The craft of gardening saw an unbounded widening of appeal in the nineteenth century. The Victorians wanted things their way, and Loudon's magazine, according to Brent Elliott,[14] was 'a forum in which

14 *Victorian Gardens*, 1986.

gardeners could discuss their work, both technical and artistic; campaign for better wages and improved working conditions; debate the political questions of the day, and the merits of cooperatives, allotments and cottage gardens.' The *Gardener's Chronicle*, launched by Joseph Paxton and John Lindley two years before Loudon's death, succeeded the *Gardener's Magazine*, becoming in its turn 'the major campaigning organ for the gardening community'. Suddenly there were an increased population, a changing social climate (not always for the better, as the novels of Charles Dickens demonstrate), the creation of new housing with garden plots in the developing suburbs around booming industrial towns, and the appearance of 'villas' where the leisured classes could cultivate their shrubberies and flower-beds with the help of paid staff. There were flower shows for everyone, new nurseries propagated the staggering influx of 'new' plants brought back to England by plant hunters in China, Japan and the Americas, and the bedding-out of such exotics as geraniums (pelargoniums), cannas and calceolarias was the order the day.

There were fads and fashions beyond the bedding out of tender exotics: ferns had their moment; rockeries were constructed with (as we would see it now) more money than taste; and societies were inaugurated to foster all manner of decorative plants. A new hunger for gardening books showed itself, and among the authors whose names have stayed fresh are Shirley Hibberd (acquainting the middle classes with his 'rustic adornments for homes of taste'); Canon Henry Ellacombe (whose gardening column in the *Manchester Guardian*, written at his parsonage in Bitton, near Bristol, is preserved in the book *In a Gloucestershire Garden*); Dean Hole of Rochester (a passionate rosarian); and those two towering figures of the late Victorian horticultural stage, who went on to dominate more than three decades of gardening in the new century, William Robinson and Gertrude Jekyll.

Born in 1838 and 1843 respectively, both Robinson and Jekyll (it rhymes with treacle) exercised an overwhelming influence, which still colours the way many people garden today. Both wrote books which remain in print, and both lived on into the 1930s and saw the results of the wholesale depletion of gardening labour brought about by the slaughter of young men in the First World War.

William Robinson was an Irishman with, according to some sources, an irascible temperament. He came to London in 1861 (to join the staff of the Royal Botanic Garden in Regent's Park), and through determination and an entrepreneurial spirit not untypical of the age he quickly mastered the publishing business, bringing out several periodicals and two classic books, *The Wild Garden* (1870) and the hugely popular *English Flower Garden* (1883). Robinson is erroneously remembered as a champion of garden informality (he clashed literary swords, notably, with Sir Reginald Blomfield, author of *The Formal Garden*), and a railer against island flower-beds shaped in 'crescents and kidneys, like flying bats or bubbling tadpoles, commingled butterflies and leeches, stars and sausages, hearts and commas, monograms and maggots';[15] but a close reading of his books saves him from the reputation of someone keen only on an unconfined style of 'natural' gardening, showing him to be sympathetic towards formal designs in those parts of the garden nearest the house. In 1885 William Robinson bought Gravetye Manor, near East Grinstead in Sussex (now a country-house hotel proud of its Robinsonian connections, where both formal and less formal parts of the garden are maintained to a high standard), and from there, well into old age, he issued his diktats on the horticultural fashions of the day.

Miss Jekyll, as she continues to be known in genteel quarters, was no less hard-working. She had a spell of editing Robinson's magazine *The Garden* (now incorporated into the extant monthly *Homes and Gardens*), but her name – and, largely, her reputation – is most closely linked with the architect Sir Edwin Lutyens (a friend with whom she collaborated on many garden designs) and a string of cherished books including *Wood and Garden* (1899), *Lilies for English Gardens* and *Wall and Water Gardens* (1901), and *Children and Gardens* and *Colour in the Flower Garden* (1908).

15 Quoted by Tyler Whittle in *Common or Garden*, 1969.

Some of Gertrude Jekyll's writing has an imperious tone ('I have no patience with slovenly planting. I like to have the ground prepared some months in advance . . .');[16] but she could also speak softly about plants, with an intimacy which indicated that there was dirt under her fingernails:

It must have been quite forty-five years ago, sometime in the early seventies of the last century [the 1870s], that I came upon a bunch-flowered Primrose in a cottage garden. I was familiar with the old laced Polyanthus, and had seen some large flowered ones of reddish colouring; but one of a pale primrose colour – something between that and white – was new to me, and I secured the plant. The next year, from some other source, came a yellowish one, much of the same character. They were of a quality that would now be thought very poor, but they were allowed to seed, and among the seedlings some of the best were kept. Gradually, from yearly selection, the quality improved, and, as the grower's judgment became more critical, so more and more of the less satisfactory Primroses were discarded. It was an immense pleasure, as the years went on, to see the coming of some new type or some new degree of colouring, and to watch for the strengthening of some desired quality.[17]

She could also be succinct: 'A dahlia's first duty in life is to flaunt and to swagger . . . and on no account to hang its head';[18] she was also, in today's parlance, streetwise: 'It is a sign of careful gardening and good upbringing, when the little boys of a family are seen on the roads with old shovels and little improvised hand-carts, collecting horse-droppings.'[19]

If Robinson and Jekyll were the king and queen of English gardening at this time, they must not be allowed to overshadow the *prince*, a man who, almost reaching his hundredth birthday, died as recently as 1954. He was Edward Augustus Bowles – gardener, botanical artist and author of distinguished monographs on narcissi and crocuses and

16 *Wood and Garden*, 1899.
17 *A Gardener's Testament*, 1937.
18 *Wood and Garden*, 1899.
19 *Old West Surrey*, 1904.

of three highly readable books about his own garden at Myddleton House, Enfield, north London.

Until the appearance of Jane Loudon and Gertrude Jekyll, women as garden writers were largely eclipsed by men (although we cannot always be sure of Anon's gender). Their presence, however, was always to be found where land was under cultivation, although William Coles, writing in 1656, prescribed a cossetted activity ('Gentlewomen if the ground be not too wet may doe themselves much good by kneeling upon a Cushion and weeding'). Sandra Raphael[20] reminds us that 'Early in the eighteenth century the Duchess of Beaufort was remarkable for her skill in cultivating the rare plants in her collection', and that Queen Charlotte, wife of George III, was an influential patron of gardening who helped to enrich the botanic gardens at Kew. Nor should we forget Eleanor Butler and Sarah Ponsonby, 'the ladies of Llangollen', who in 1778 'eloped' from Ireland and later set up house at Plas Newydd in North Wales, where their garden was visited by such swells of the day as Lady Caroline Lamb, Josiah Wedgwood, Wordsworth and Southey. And spanning the nineteenth and twentieth centuries there was the somehow sad figure of Ellen Willmott, an elegant, wealthy spinster whose obsession for making gardens (at Warley Place in Essex, and on the Riviera in France and Italy) led almost to pitiful bankruptcy.

Jane Loudon (b. 1807) was, as we have seen, a pioneer, but she was not alone. Theresa (Mrs C. W.) Earle, whose *Pot-Pourri from a Surrey Garden* must have found its way to India in many a memsahib's portmanteau, was only thirty years her junior: Mrs Alice Morse Earle (no relation) published *Old Time Gardens* in New York in 1901; Alicia Amherst wrote *A History of Gardening in England* in 1895; another American, the novelist Edith Wharton, wrote a book about Italian villas and their gardens in 1904; the Hon. Mrs E. V. Boyle's chatty *Days and Hours in a Garden* came out in 1884; Vita Sackville-West,

20 *The Oxford Companion to Gardens*, 1986.

whose name is to be found throughout most of twentieth-century gardening, lived the first nine years of her life while Queen Victoria was on the throne.

But it is on the spines of twentieth-century gardening books that women's names have become paramount. Beth Chatto, Marion Cran, Dame Sylvia Crowe, Margery Fish, Penelope Hobhouse, Dawn MacLeod, Mirabel Osler, Frances Perry, Anne Scott-James, Jane Taylor and Rosemary Verey are but a few of the authors who have found a wide and appreciative readership, often extending beyond their native Britain. Deborah Kellaway's *Virago Book of Women Gardeners* (1995) includes passages from books and articles by over 100 women, while the galaxy of twentieth-century male garden writers is no less brilliant, including Robin Lane Fox, Arthur Hellyer, Anthony Huxley, A. T. Johnson, John Kelly, Stephen Lacey, Roy Lancaster, Christopher Lloyd, Russell Page, Patrick Synge and Graham Stuart Thomas.

Vita, as she is familiarly known to anyone with an interest in gardening, was born in 1892. After 'losing' her ancestral home, Knole, and making her first garden at Long Barn near Sevenoaks, she bought Sissinghurst Castle in Kent, and with her husband, Harold Nicolson, created there what was to become one of the western world's best-loved gardens.

Despite the 'advancement' in garden design in postwar years, it is at Sissinghurst (and at Hidcote Manor, where Lawrence Johnston, a generation before the Nicolsons, fashioned his own bit of Paradise on a blowy Cotswold escarpment) that the history of garden-making in this country completes one of its circles. Harold and Vita's *allées*, nut walk, garden rooms and flowery orchard are not so very different from the enclosed monastic gardens and bespangled meads of earlier times, and, although the number and variety of plants have increased beyond any possible expectation, it is pleasing to think that in the face of ever-changing garden trends, and (in this century alone) two world wars and an explosion of electronic technology, the likes of Andrew Marvell – should he care at all to reincarnate himself in this age of post-modernism and inter-planetary exploration – might still find somewhere for his 'green Thought in a green Shade'.

DAVID WHEELER
Bryan's Ground
Stapleton (nr Presteigne), Herefordshire

CHAPTER ONE

Thoughts on the Garden

THE GARDEN

How vainly men themselves amaze
To win the Palm, the Oke, or Bayes;
And their uncessant Labours see
Crown'd from some single Herb or Tree,
Whose short and narrow verged Shade
Does prudently their Toyles upbraid;
While all Flow'rs and all Trees do close
To wave the Garlands of repose.

Fair quiet, have I found thee here,
And Innocence thy Sister dear!
Mistaken long, I sought you then
In busie Companies of Men.
Your sacred Plants, if here below,
Only among the Plants will grow.
Society is all but rude,
To this delicious Solitude.

No white nor red was ever seen
So am'rous as this lovely green.
Fond Lovers, cruel as their Flame,
Cut in these Trees their Mistress name.
Little, Alas, they know, or heed,
How far these Beauties Hers exceed!
Fair Trees! where s'eer your barkes I wound,
No Name shall but your own be found.

When we have run our Passions heat,
Love hither makes his best retreat.
The *Gods*, that mortal Beauty chase,
Still in a Tree did end their race.
Apollo hunted *Daphne* so,
Only that She might Laurel grow.
And *Pan* did after *Syrinx* speed,
Not as a Nymph, but for a Reed.

What wond'rous Life in this I lead!
Ripe Apples drop about my head;
The Luscious Clusters of the Vine
Upon my Mouth do crush their Wine;
The Nectaren, and curious Peach,
Into my hands themselves do reach;
Stumbling on Melons, as I pass,
Insnar'd with Flow'rs, I fall on Grass.

Mean while the Mind, from Pleasure less,
Withdraws into its happiness:
The Mind, that Ocean where each kind
Does streight its own resemblance find;
Yet it creates, transcending these,
Far other Worlds, and other Seas;
Annihilating all that's made
To a green Thought in a green Shade.

Here at the Fountains sliding foot,
Or at some Fruit-trees mossy root,
Casting the Bodies Vest aside,
My Soul into the boughs does glide:
There like a Bird it sits, and sings,
Then whets, and combs its silver Wings;
And, till prepar'd for longer flight,
Waves in its Plumes the various Light.

Such was that happy Garden-state,
While Man there walk'd without a Mate:
After a Place so pure, and sweet,
What other Help could yet be meet!
But 'twas beyond a Mortal's share
To wander solitary there:
Two Paradises 'twere in one
To live in Paradise alone.

How well the skilful Gardner drew
Of flow'rs and herbes this Dial new;
Where from above the milder Sun
Does through a fragrant Zodiack run;
And, as it works, th'industrious Bee
Computes its time as well as we.
How could such sweet and wholesome Hours
Be reckon'd but with herbs and flow'rs!

ANDREW MARVELL, *c.* 1651

I asked a schoolboy, in the sweet summer-tide, 'what he thought a garden was for?' and he said, *Strawberries.* His younger sister suggested *Croquet,* and the elder *Garden-parties.* The brother from Oxford made a prompt declaration in favour of *Lawn Tennis and Cigarettes,* but he was rebuked by a solemn senior, who wore spectacles, and more back hair than is usual with males, and was told that 'a garden was designed for botanical research, and for the classification of plants'. He was about to demonstrate the differences between the *Acoty-* and the *Monocoty-ledonous* divisions, when the collegian remembered an engagement elsewhere.

I repeated my question to a middle-aged nymph, who wore a feathered hat of noble proportions over a loose green tunic with a silver belt, and she replied, with a rapturous disdain of the ignorance which presumed to ask – 'What is a garden for? For the soul, sir, for the soul of the poet! For visions of the invisible, for grasping the intangible, for hearing the inaudible, for exaltations' (she raised her hands, and stood tiptoe, like jocund day upon the misty mountain top, as though she would soar into space) 'above the miserable dulness of common life into the splendid regions of imagination and romance.' I ventured to suggest that she would have to do a large amount of soaring before she met with anything more beautiful than the flowers, or sweeter than the nightingale's note; but the flighty one still wished to fly.

A capacious gentleman informed me that nothing in horticulture touched him so sensibly as green peas and new potatoes, and he spoke with so much cheerful candour that I could not be angry; but my indignation was roused by a morose millionaire, when he declared that of all his expenses he grudged most the outlay on his confounded garden.

Dejected, I sought solace from certain ladies and gentlemen, who had expressed in my hearing their devoted love of flowers. They were but miserable comforters. Their devotion was superficial, their homage conventional: there was no heart in their worship. I met with many who held flowers in high estimation, not for their own sake, not for the loveliness and perfect beauty of their colour, their fragrance, and their form, not because even Solomon in all his glory was not arrayed like one of these, but because they were the most effective decorations of their window-sills, apartments, and tables, and the most becoming embellishments for their own personal display. I found gentlemen who restricted their enthusiasm to one class of plants, ignoring all the rest; and even in this their valuation was regulated by the rarity and the cost of the flower. 'I can assure you, my dear sir,' they said, 'that there is only one other specimen in the country, and that the happy possessor is my friend, Lord Lombard.' And I shall never forget the disastrous results which followed, when I informed one of these would-be monopolists, that I knew a third party, who had duplicates. He favoured me with a scowl of mingled disgust and doubt, sulked during the remainder of our interview, and became my bitter enemy for life.

> Such men as he are never at heart's ease,
> While they behold a greater than themselves.

Others were quite as exclusive, but with a difference of intention. They not only desired to possess, but that the public should know that they possessed, something out of the common; and from their love of renown, or their '*sacra auri fames*', they competed for the prizes which were awarded to their favourite flower. It seemed to me that they derived much more gratification from the cups and stakes than from the horses, who had won the race.

The unkindest cut of all, so common that it makes one callous, comes from those visitors who 'would be so delighted to see our garden!' and they come and see, and forget to be delighted. They admire the old city walls which surround it, they like to hear the cawing of the rooks, they are pleased with the sun-dial and the garden-chairs, but as for horticulture they might as well be in Piccadilly! They would be more attracted by the fruit in Solomon's shop than by all the flowers in the border. I heard a lady speaking to her companion of 'the most perfect gem she had ever seen', and when, supposing that reference was made to some exquisite novelty in plants, I inquired the name and habitation, I was informed that the subject under discussion was 'Isabel's new baby'! 'Ladies,' I remarked, with a courteous but scathing satire, 'I have been a baby myself, and am now a proprietor, but I am constrained to inform you that this is a private, and not a nursery, garden.'

Thus disappointed, deceived, disheartened, I began to fear that my intense love of a garden might be a mere hallucination, an idiosyncrasy, a want of manliness, a softening of the brain. Nevertheless I persevered in my inquiries, until I found that which I sought – the sympathy of an enthusiasm as hearty as my own, a brotherhood and a sisterhood, who, amid all the ignorance and pretence of which I have given examples, were devoted to the culture of flowers, and enjoyed from this occupation a large portion of the happiness, which is the purest and the surest we can know on earth, the happiness of Home.

S. REYNOLDS HOLE, 1899

If you have anything like a visual memory – the kind of memory that obligingly turns over pages of pleasantly coloured pictures for you on the slightest request, or even unbidden, but, modestly, blurs their lines when you would set them down on paper – there is probably in its collection some unforgettably vivid scene of flowers growing as you have never been able to grow them in your garden: primroses, perhaps in an oak-coppice, or gorse in full flower, or ling and bell-heather stretching up to a well-known skyline. The psychologists will probably tell you that all your efforts in the garden are influenced in some way by these vivid memories: they will also be careful to explain that they are even more influenced by something which you think you have entirely forgotten, and are keeping close bolted in an even more remote corner of your mind. It may be so, and it may not: it does not really matter very much, unless you are anxious to discover whatever made you take up gardening at all – a point on which I find most gardeners are slightly evasive. Confidences are rather out of fashion to-day compared with fifty years ago: emotions are considered to be shy-making, and anything which is not of demonstrably practical value is to be deprecated. Your good gardener will admit that his garden is an extravagance on which he spends more time and money than he cares to think about, and that the results are incommensurate with his expenditure – particularly in the kitchen garden, against which, as a sop to the decencies, he probably has a lurking grudge – but the proverbial wild horses will not make him talk freely about the pleasurable emotions which his gardening gives him, or the deep-seated needs which it satisfies: and, equally inhibited yourself, and probably even more extravagant, you will not press him.

HUMPHREY DENHAM, 1940

If the man who makes two blades of grass grow where one grew before deserves well of his fellows, surely he who turns plainness into beauty should be put upon a pedestal for worship, and, better still, for imitation.

FREDERICK EDEN, 1903

If only one were as good a gardener in practice as one is in theory, what a garden one would create!

VITA SACKVILLE-WEST, 1937

The love of dirt is among the earliest of passions, as it is the latest. Mud-pies gratify one of our first and best instincts. So long as we are dirty, we are pure. Fondness for the ground comes back to a man after he has run the round of pleasure and business, eaten dirt, and sown wild-oats, drifted about the world, and taken the wind of all its moods. The love of digging in the ground (or of looking on while he pays another to dig) is as sure to come back to him as he is sure, at last, to go under the ground, and stay there. To own a bit of ground, to scratch it with a hoe, to plant seeds, and watch their renewal of life, – this is the commonest delight of the race, the most satisfactory thing a man can do.

CHARLES DUDLEY WARNER, 1876

The garden, like every other work of art, should have a climax, which may be emphasized by moving water or by reflection, by higher light and deeper shadow, by colour, elaboration in design, or by sculpture which adds the charm of personality; and whatever this focus of interest may be, whether it stirs the emotion of sublimity by the prospect of mountain amphitheatre or plunging cataract or measureless infinity of plain, the emotion of beauty by immemorial cypresses, by lake or river valley, blue pool or fountain basin, it must be presented with what is known as 'economy of the recipient's attention'; that is to say, without the addition of features which disturb or detract. If a picture be complete, everything that is added is something taken away. The Law of Novelty will teach us not to fritter away the effect in half glimpses, but to deliver it with unsullied freshness, like a knock-down blow straight between the eyes; for it is a fundamental principle that no second occurrence of any great stimulus can ever be fully equal to the first. The power of novelty to 'quicken observation, sharpen sensation and exalt sentiment' is as marked as that of familiarity to throw a veil over ugliness. Surprise may add to novelty the shock of contradiction, and if care has been taken to make the expectation less than the reality, we shall have the added thrill of wonder. Indeed, in every great garden there should be some element of wonder or surprise, if only to make recollection more vivid.

SIR GEORGE SITWELL, 1909

By this time it was getting dark, and it was very pleasant in Grove Street, where most of the good people had just watered their little gardens, and brought out the sweetness of the mignonette. Mr Cavendish was not sentimental, but still the hour was not without its influence; and when he looked at the lights that began to appear in the parlour windows, and breathed in the odours from the little gardens, it is not to be denied that he asked himself for a moment what was the good through all this bother and vexation, and whether love in a cottage, with a little garden full of mignonette and a tolerable amount of comfort within, was not, after all, a great deal more reasonable than it looked at first sight.

MRS OLIPHANT, 1866

. . . inside the garden, we can construct a country of our own. Several old trees, a considerable variety of level, several well-grown hedges to divide our garden into provinces, a good extent of old well-set turf, and thickets of shrubs and evergreens to be cut into and cleared at the new owner's pleasure, are the qualities to be sought for in your chosen land. Nothing is more delightful than a succession of small lawns, opening one out of the other through tall hedges; these have all the charm of the old bowling-green repeated, do not require the labour of many trimmers, and afford a series of changes. You must have much lawn against the early summer, so as to have a great field of daisies, the year's morning frost; as you must have a wood of lilacs, to enjoy to the full the period of their blossoming. Hawthorn is another of the Spring's ingredients; but it is even best to have a rough public lane at one side of your enclosure which, at the right season, shall become an avenue of bloom and odour. The old flowers are the best and should grow carelessly in corners. Indeed, the ideal fortune is to find an old garden, once very richly cared for, since sunk into neglect, and to tend, not repair, that neglect; it will thus have a smack of nature and wildness which skilful dispositions cannot overtake. The gardener should be an idler, and have a gross partiality to the kitchen plots: an eager or toilful gardener misbecomes the garden landscape; a tasteful gardener will be ever meddling, will keep the borders raw, and take the bloom off nature. Close adjoining, if you are in the south, an olive-yard, if in the north, a swarded apple-orchard reaching to the stream, completes your miniature domain; but this is perhaps best entered through a door in the high fruit-wall; so that you close the door behind you on your sunny plots, your hedges and evergreen jungle, when you go down to watch the apples falling in the pool. It is a golden maxim to cultivate the garden for the nose, and the eyes will take care of themselves. Nor must the ear be forgotten: without birds a garden is a prison-yard.

ROBERT LOUIS STEVENSON, 1879 (pub. 1905)

I have always wanted a garden of my own. Other people's gardens are all very well, but the visitor never sees them at their best. He comes down in June, perhaps, and says something polite about the roses. 'You ought to have seen them last year,' says his host disparagingly, and the visitor represses with difficulty the retort, 'You ought to have asked me down to see them last year.' Or, perhaps, he comes down in August, and lingers for a moment beneath the fig-tree. 'Poor show of figs,' says the host, 'I don't know what's happened to them. Now we had a record crop of raspberries. Never seen them so plentiful before.' And the visitor has to console himself with the thought of the raspberries which he has never seen, and will probably miss again next year. It is not very comforting.

Give me, therefore, a garden of my own. Let me grow my own flowers, and watch over them from seedhood to senility. Then shall I miss nothing of their glory, and when visitors come I can impress them with my stories of the wonderful show of groundsel which we had last year.

For the moment I am contenting myself with groundsel. To judge by the present state of the garden, the last owner must have prided himself chiefly on his splendid show of canaries. Indeed, it would not surprise me to hear that he referred to his garden as 'the back-yard'. This would take the heart out of anything which was trying to flower there, and it is only natural that, with the exception of the three groundsel beds, the garden is now a wilderness. Perhaps 'wilderness' gives you a misleading impression of space, the actual size of the pleasaunce being about two hollyhocks by one, but it is the correct word to describe the air of neglect which hangs over the place. However, I am going to alter all that.

With a garden of this size, though, one has to be careful. One cannot decide lightly upon a croquet-lawn here, an orchard there, and a rockery in the corner; one has to go all out for the one particular thing, whether it is the last hoop and the stick of a croquet-lawn, a mulberry-tree, or an herbaceous border. Which do we want most – a fruit garden, a flower garden, or a water garden? Sometimes I think fondly of a water garden, with a few perennial gold-fish flashing swiftly across it, and ourselves walking idly by the margin and pointing them out to our visitors; and then I realize sadly that, by the time an adequate margin has been provided for ourselves and our visitors, there will be no room left for the gold-fish.

At the back of my garden I have a high brick wall. To whom the bricks actually belong I cannot say, but at any rate I own the surface rights on this side of it. One of my ideas is to treat it as the back cloth of a stage, and paint a vista on it. A long avenue of immemorial elms, leading up to a gardener's lodge at the top of the wall – I mean at the end of the avenue – might create a pleasing impression. My workroom leads out into the garden, and I have a feeling that, if the door of this room were opened, and then hastily closed again on the plea that I mustn't be disturbed, a visitor might obtain such a glimpse of the avenue and the gardener's lodge as would convince him that I had come into property. He might even make an offer for the estate, if he were set upon a country house in the heart of London.

But you have probably guessed already the difficulty in the way of my vista. The back wall extends into the gardens of the householders on each side of me. They might refuse to co-operate with me; they might insist on retaining the blank ugliness of their walls, or endeavouring (as they endeavour now, I believe) to grow some unenterprising creeper up them; with the result that my vista would fail to create the necessary illusion when looked at from the side. This would mean that our guests would have to remain in one position, and that even in this position they would have to stand to attention – a state of things which might mar their enjoyment of our hospitality. Until, then, our neighbours give me a free hand with their segments of the wall, the vista must remain a beautiful dream.

However, there are other possibilities. Since there is no room in the garden for a watchdog *and* a garden, it might be a good idea to paint a phosphorescent and terrifying watchdog on the wall. Perhaps a watch-lion would be even more terrifying – and, presumably, just as easy to paint. Any burglar would be deterred if he came across a lion suddenly in the back garden. One way or another, it should be possible to have something a little more interesting than mere bricks at the end of the estate.

And if the worst comes to the worst – if it is found that no flowers (other than groundsel) will flourish in my garden, owing to lack of soil or lack of sun – then the flowers must be painted on the walls. This would have its advantages, for we should waste no time over the early and uninteresting stages of the plant, but depict it at once in its full glory. And we should keep our garden up to date. When delphiniums went out of season, we should rub them out and give you chrysanthemums; and if an untimely storm uprooted the chrysanthemums, in an hour or two we should have a wonderful show of dahlias to take their place. And we should still have the floor-space free for a sundial, or – if you insist on exercise – for the last hoop and the stick of a full-sized croquet-lawn.

A. A. MILNE, 1920

I have no garden of my own and what I have learned has been from the years of working on other people's ground. But I allow myself day-dreams that I one day mean to have.

I hope that it will be on land which is neither chalky nor too acid, with soil to which I can add peat or lime. I hope, too, for soil which is not heavy clay that takes years of back-breaking labour to lighten, nor a hot dry sand whose thirst for water and manure I could never assuage.

It must be a small garden and a simple one; one man's work, mine perhaps, and in any case not so large as to need an armoury of mechanical devices or a full-time mechanic to keep these in running order. However good the soil, the first thing I shall do will be to make two enclosures, for compost, each five feet wide by ten feet long and walled in to a height of three feet. Into these will go everything that in a few months will ensure me a regular supply of rich black humus, since I know no better way of having a garden relatively free from pests and diseases. There are limits to the time, trouble and money I am prepared to spend on spraying my garden with chemical preparations which can so easily destroy nature's subtle balances and, by eliminating one pest, fatally leave the door open for others.

Once I have made this, the garden's future larder, I can start to consider the site. First of all I should say that I plan to make my garden in England since, all things considered, I do not know a better country. I would rather start with an old garden, however badly arranged and however neglected, since a few mature trees, an old wall and even a few square yards of good soil will give me the advantage of a twenty year start, all the more so as I shall be so late a starter. First I shall take out all the rubbish, elder bushes, nettle beds and any trees which are ugly and misshapen or too crowded. I shall thin out old shrubberies without pity and prune back any specimens which I may wish to keep and, later perhaps, transplant. All soft green rubbish will go to the compost heaps, the rest I shall burn and save the wood ash where I can. Only when I have cleaned the garden and ridded it of everything I know I will not want, shall I make a careful survey on paper as a basis for an eventual plan.

My garden will be very simple. There will be no herbaceous borders, no rose garden and no complicated formal layout with all the bedding out, edge trimming and staking which these forms of gardening involve. A ground-floor room in the house or a converted outbuilding will be my workroom, part library for garden books and catalogues, part studio for drawing board and painting materials, part tool shed for all the small tools, string, raffia, tins of saved seed and all the odd extensions and aids to the gardener's two hands.

I see my workroom with one wall all window, and below it a wide work table running its whole length with a place to draw and a place to write. Walls will be whitewashed, the floor of brick, there will be a fireplace and chairs for talk, and at least one wall lined with books.

This room will open south or westwards on to my working garden and I shall design this just like one of those black-japanned tin boxes of water-colours. I see this working garden as a rectangular space as large as I can afford and manage. With luck or good management or both, high walls will protect it to east and north and I shall enclose the other two sides with a low wall or a five feet yew or box hedge. Under the walls I shall make a three or four feet wide bed for climbing plants and others which like the warmth and shelter that a wall at their back will give them. I shall then divide the rest into small beds, perhaps four and a half feet or five feet square, separated by eighteen inch paths of bricks on edge or stone or even pre-cast cement flags of a good texture and colour and set, like the bricks or stone, in a weak cement mixture so that no weeds can grow. The number of beds and their exact dimensions will depend on the area of my ground but they must be small enough to be easily accessible from the paths surrounding them. In working out my simple criss-cross pattern I shall surely find it useful to have a few beds as double units, nine feet by four foot six or ten feet by five feet. In fact this garden will closely resemble the 'system garden' of an old botanical garden whose small beds are each devoted to growing numbers of the different plant families. I shall use this garden as paint box, palette and canvas, and in it I shall try out plants for their flower colour, texture of foliage and habit of growth. In some beds I shall set out seedlings for selection, in others bulbs, in others plants combined for essays in colour. Each bed will be autonomous, its own small world in which plants will grow to teach me more about their æsthetic possibilities and their cultural likes and dislikes. I shall make no attempt at a general effect, for this will be my personal vegetable museum, my art gallery of natural forms, a trial ground from which I will always learn.

RUSSELL PAGE, 1962

In every year there are days between winter and spring which rightly belong to neither; days when the round of the seasons seems to be at a standstill, as though the inner impulse which held on visibly enough through the worst of the hard weather had failed just when it should begin to quicken towards the first of the better times. There are evenings in an interregnum of this kind when the world is as nearly unlovely as it ever can be under natural conditions: the air cold with a sodden chill that bites worse than frost; land and sky wrapped in a dim cloud without form or motion; the year altogether at its worst, foul from the winter, frost-bitten, flood-swept, sunk in a sapless lethargy when it is more than time for the stirrings of the yearly miracle of repair. To any countryman these dead seasons offer a fair exercise in the spirit of hopefulness and faith in to-morrow; to a man who cares for a garden they are among the severest of the trials so wholesomely afforded by the garden year. The gardener is not waiting only for softer air and whiter sun; he has the almanack and his old diaries to add stings to his impatience; seed-time is overdue; the irreparable arrears are being scored against him; it is ten days since the date when the first venture of early potatoes, or the peas on which something of his reputation rests, should have been got in under the accustomed south wall. He paces the walks disconsolate; the soil of beds and quarters, waterlogged, or reduced by alternate small frosts and thaws to an unapproachable slime, needs a blustering week of sun to make it fit for the spade. The proper outdoor works reserved for hard weather, the stick-cutting and leaf-gathering, and the fireside businesses of rotation plans and seed lists are all done to a painful completeness; there is nothing left but again to turn over the thrice-turned compost heaps, to bank up the smouldering weed fire, and to entertain the reflections which the enforced holiday provides.

At the outset, as the spade jars on the frozen crust of the turf stack and the fingers tingle to the nipping air, these meditations will probably be concerned with the present iron age; with the worst season ever known for mildew, with the hopelessness of getting a decent seed bed, with the hard-hit tea roses, with the sweeping havoc, spite of netting and scaring devices, which the tits and finches are making among the buds of the small fruit and pears. From the present the mind works back to past seasons, recalling the old losses and hard times – that six weeks of black frost which cut down the laurels; the wet, clogging snow which split the great cedar and wrecked half the shrubberies; the drowned summers and brazen droughts. The tale of failures, all the seeds that never came up, the buddings that missed, lie heavy on the temper. But after half an hour's shovelling, a subtle change begins to work; the air does not bite so rawly, the turf and the leaf mould are in noble trim, a whiff of smoke from the weed-smother touches some associative sensory with an odd, irrelevant pleasure. The calamities of the past begin to give way to its triumphs and its white days; and the transition is eased by a not unpleasing pathos in the contrast with the present degeneracy. There come to mind the year which enjoyed green peas from May to late November, the seedling pansies which amazed the village flower-show, the incredible circumference of those Catillac pears. And when at length the compost-stacks are all turned inside out and orderly piled up again, there comes a time for sitting down on the edge of a handy cucumber-frame or wheelbarrow handle, for breathing long steady draughts of the air which now holds no more than a fine stringent freshness, charged with the rich wholesomeness of disturbed turf and mould. For a few minutes' meditation the time is not so ill, this quiet-fallen end of day when the smoke from the fire drifts and hangs ghost-like about the dark ground, and when the misty dusk, confusing landmarks and boundaries, affords a vague background available for all sorts of garden visions, from the suburban plot where a prentice hand may to his perpetual advantage learn to respect the London clay, down to the utmost Hesperides of one's good dreams, where comes neither flood nor drought, where there are spells against *Phytophthora infestans* and where the wireworm and the codlin-moth are unknown. It is the humour of the latter vision which prevails; with the sense of virtuous content produced by stretched muscles and opened lungs comes back the consciousness of the golden year which is never far below the

gardener's horizon. Things must take a turn; the balance of likeli-
hoods proclaims a full measure of sunshine, after last year's dripping
skies; this spring the April frosts may be the exception to prove the
rule. Across the drenched borders the eye catches the glimmer of
blue-green tufts, or ranks of shadowy sword-points, while the mind
sees the May sunlight full on the iris flowers, the carnations stirring to
the warm wind in July. And if it should seem too hard to believe in
that transformation from frosted leaves and dry rhizomes, here are the
daffodil-sheaths pushing through the mould, no wintered survival, but
new life stirring in the teeth of the improbabilities.

The wise man will always recognize the value of the compulsory
holidays which, almost alone of the great primary arts, gardening
enjoys: – gardening, rather than husbandry in general, in England, at
least, and under the present conditions, because with us the gardener is
left sole heir to nearly all the original blessings of the tiller of the soil.
The modern British farmer, whether he hold a thousand acres of corn
land in the Shires, or only a hundred of Sussex thistles, has in the main
lost the mark of virtues once peculiar: – leisurely work, waiting on the
seasons, breadth of view and security of temper based on the immemorial treaty between tolerable soil and rational labour. He has no
time to catch his balance in the classic attitude of equanimity; the
thought of income tax and of the requirements of the Board of
Agriculture prevents the effectual grace; the least enlightened small
farmer is nowadays rather too much of a chemist and far too much of
an engineer to admit the true vocation. It is the working gardener of
the right breed, one that is not above taking the risk of lumbago on his
own back, yet has the temperament and the opportunity for a reflect-
ive humour, to whom that saying of Xenophon rightly applies, that the
husbandman's art is the easiest of all to learn, and the pleasantest to
exercise, and is most profitable for the body, and least of all robs the
soul of the leisure needful for properly attending to our friends and
country . . .

Whether he remember his Xenophon or not, the contemplative
gardener may well reflect during his ten minutes of twilight leisure that
he is the right representative of an ancient art, once held the mother

and nurse of all the rest. Whatsoever expedients may be devised by inquirers into national degeneration, he has done his part to uphold the standard in the earlier way. If by chance the old judgment should be found right after all, that there are some businesses which ruin the body and shatter the soul, he for one is clear of the belief that a whole nation may profitably devote itself to these alone. If all the world turn to blast-furnaces and coal-pits, he and his kind will scratch out their little plots amongst the slag heaps, and revert perennially instinctive survivals, to their primitive craft.

And thus musing and magnifying his office, he feels again that the Great Year is not so very far below the dark edge of the world. Spite of longer session than he intended on the cucumber-frame, the air seems to have lost its sting; and as he picks up his tools for the night he catches sight, high up through a momentary gap in the mist, of a faint flush of sunset on high-streaming cloud that moves – that certainly moves, by observation taken against the cedar branches overhead – from the South once more.

ANON, before 1914

Gardens are a paradox. They reflect their owners; they are totally dependent; and yet in no time at all they are breathing with their own lungs, growing at their own pace, behaving with either wilful disregard or subjugation. Subjugated gardens abound, and I can see why. Unless discipline is maintained from the moment the spirit-level is laid across the earth, you are nurturing a vast, tactile, heavy-scented siren which will keep you forever in its thrall. Happy and doting, you never stop planning for its well-being. Though there may be little to be said for huge bulks of gardening, to be without one is utter deprivation. You cut yourself off from a whole spectrum of sensory ferment whose intensity you can have no idea exists until you walk in it very early one spring morning, when the light lies like a membrane over the land and tulips are leavening the orchard with their vigour. Or when you sit fazed in the broody light of a mid-summer afternoon to the murmur of bees and a distant hay cutter, and discover then with what brilliance and with what almost unpremeditated intuition you have created something of overwhelming beauty. How easy to become deflected by the marbled leaves of cyclamen with their brittle stems and flighty petals, or turn back one more time to the sight of bright tulips spread-eagled under a leaning tree, or succumb to a basking *Rosa paulii* just outside the door, lying like a beached whale, exuding fragrance.

MIRABEL OSLER, 1990

There are some pretty glum faces around here. Glum? No, re-signed to the frailty of the human hand, perhaps. It is the season of storms and failed resolutions, both of which bode well for the monster lying in wait at the door – the garden. I've always considered 'the garden' to be an inappropriate description of any acres I have toiled upon. I've often toyed with the idea of calling it – even bad-naming it – 'my green conscience', but then I might be mistaken for the guardian of the planet, which I am not. Our monster, you must under-stand, is the sort of creature that revels in matted hair and dirty fingernails. If it finds a puddle it will jump in it, and then roll about in it. It also has little regard for order, and no respect for its begetters. It hates its uniform and behaves badly when there are visitors. It will shout 'shan't' at the top of its voice and throw its toys about – some-times the furniture is upended. We like to think this a phase it's going through, but no; every winter it's the same. And then, just when thoughts of abandonment flicker through one's mind, after a week of uprooting stakes, paddling in flood water, flattening fences, peeling back the roofs of potting sheds, and inviting the sheep-next-door around for a meal, the little darling totally disarms you by presenting a midwinter posy from behind its back.

TIMOTHY RIBBESFORD, 1993

And now a garden pland with nicest care
Should be my next attention to prepare;
For this I'd search the soil of different grounds
Nor small nor great should mark its homley bounds:
Between these two extreems the plan should be
Compleat throughout and large enough for me;
A strong brick wall should bound the outward fence
Where by the suns allcheering influence
Walltrees should flourish in a spreading row
And Peach and Pear in ruddy lustre glow.
A five foot bed should follow from the wall
To look compleat or save the trees withall
On which small seeds for sallading I'd sow
While curl-leaf Parsley should for edges grow.
My Garden in four quarters I'd divide
To show good taste and not a gaudy pride;
In this the middle walk should be the best
Being more to sight exposed than the rest,
At whose southend a harbour should be made
So well belov'd in summer for its shade:
For this the rose would do or jessamine
With virginbower or the sweet woodbine,

Each one of these would form exactly well
A compleat harbour both for shade or smell.
Here would I sit when leisure did agree
To view the pride of summer scenery
See the productions promis'd from my spade
While blest with liberty and cooling shade.
But now a spot should be reserv'd for flowers
That would amuse me in those vacant hours
When books and study cease their charms to bring
And Fancy sits to prune her shatterd wing,
Then is the time I'd view the flowrets eye
And all loose stragglers with scotch-mattin tye;
The borders too I'd clean with nicest care
And not one smothering weed should harbour there:
In trifling thus I should such pleasure know
As nothing but such trifles could bestow.

JOHN CLARE, 1808–9

Seeing all the Designs we form to ourselves are nothing but the pure Effects of our Fancy, we should be very deceiv'd if by Study or Reflection we did not endeavour to find out Means to render the Execution of them easie, which can no way be better effected, than by having Recourse to the most solid and obvious Rules of Art . . .

I pretend not, as several Authors have done, to confine the Lovers of a Garden of this kind to any particular Situation, as being the best and most advantagious; that would be to restrain them too much, in case they had not the good Fortune to light upon just such a Piece of Ground as I should prescribe, without which, according to that Rule, all their endeavours would be but imperfect.

In Matters of Art, the Mind ought by no means to be check'd; but after a Way is open'd for the entire Knowledge of a Thing, the Execution of it ought to be made as easie as possible.

'Tis true, a Piece of Ground that lyes sloping, and looks towards the rising Sun, is most esteem'd, because the Water runs easier from it, and the Sun at his Rising coming to dart on the Flowers in a Parterre thus situated, so enlivens them, that by means of his Heat on the Morning Dew still remaining upon them, they become much more vigorous in their Growth, than if the Parterre were in another Situation.

This Sloping, nevertheless, which is so much esteem'd, is not always necessary, especially in very dry Soils, where the stopping of the Water does rather good than harm; so that in light, sandy, lean Soils, and the like, no Regard is to be had to such a Declivity.

As for Soils that are extreamly wet, this sloping Situation must be had if possible, because the Flowers we commonly cultivate in our Gardens by no means love to have their Roots always in Water. However, if it should be a Man's Lot to have a House situated upon a Flat, I would not have him, for that Reason only, abandon his Design of making a Flower-Garden, since where the Nature of a Place will not allow us all we could wish, Art comes to our Relief, supplying our Wants to that degree, that we may easily have wherewithal to content our Desires . . .

This sloping Situation is not likewise to be consider'd in relation to stony Soils, which being easily warm'd, not only by the Heat of the Sun, but by means of the Stones retaining that Heat, they can hardly ever have Moisture enough.

Those happy Soils which naturally afford good Productions, how little soever they are cultivated, do very well either with or without this Sloping; for let the Parterre, or Ground-Plot of such a Flower-Garden, be situate where it will, all may succeed in it, provided nothing be omitted to make the Plants grow . . .

Some pretend the Vapours which arise from marshy Grounds are pernicious to Flowers, and consequently that a Flower-Garden ought, by no means, to be planted near Marshes: But I am of Opinion this is a needless Scruple, since in many of these Places Flowers may thrive as as elsewhere, provided no Pains or Care be spar'd to make them do so, and that the Parterre be not plac'd too near a Wood, whose Shade would be apt to hinder it from enjoying the free Air and Sun. Flowers have naturally no Antipathy to Moisture, in case it be immoderate; on the contrary, it makes them more lively, and last the longer.

As to the Aspects that agree best with Parterres, there are some who would impose Laws upon us concerning them, affirming that a Garden design'd for Flowers will come to little unless it be expos'd to the East, and be shelter'd from the North Winds by a Wall: But Experience teaches us every Day, there are Flower-Gardens in all Aspects which produce excellent Flowers. Thus without troubling his Head any further about preferring one Aspect before another, a Florist may succeed well enough with his Flowers, whatever Exposition his Garden lyes in, provided it be not too much in the Shade; for where the Sun comes but little, or not at all, whatever we plant never thrives well.

'Tis agreed by all, that the South Aspect is not so kindly to Flowers, which are Plants of a tender Constitution, as that of the East, and when a Man is at liberty to chuse his Parterre where he pleases, the latter is preferable to the former; but when one is oblig'd to make use of a Piece of Ground as one finds it, every Situation, as I have said before, may do well enough in cultivating Flowers . . .

FRANÇOIS GENTIL, 1706

G od *Almightie* first Planted a *Garden.* And indeed, it is the Purest of Humane pleasures. It is the Greatest Refreshment to the Spirits of Man; Without which, *Buildings* and *Pallaces* are but Grosse Handy-works: And a Man shall ever see, that when Ages grow to Civility and Elegancie, Men come to *Build Stately*, sooner then to *Garden Finely*: As if *Gardening* were the Greater Perfection . . .

For *Gardens* ... the Contents, ought not well to be, under *Thirty Acres of Ground*; And to be divided into three Parts: A *Greene* in the Entrance; A *Heath* or *Desart* in the Going forth; And the *Maine Garden* in the midst; Besides *Alleys*, on both Sides. And I like well, that Foure Acres of Ground, be assigned to the *Greene*; Six to the *Heath*; Foure and Foure to either *Side*; And Twelve to the *Maine Garden*. The Greene hath two pleasures; The one, because nothing is more Pleasant to the Eye, then Greene Grasse kept finely shorne; The other, because it will give you a faire Alley in the midst, by which you may go in front upon a *Stately Hedge*, which is to inclose the *Garden*. But, because the Alley will be long, and in great Heat of the Yeare, or Day, you ought not to buy the shade in the *Garden*, by Going in the Sunne thorow the *Greene*, therefore you are, of either *Side* the *Greene*, to Plant a *Covert Alley*, upon Carpenters Worke, about Twelve Foot in Height, by which you may goe in Shade, into the *Garden*. As for the Making of *Knots*, or *Figures*, with *Divers Coloured Earths*, that they may lie under the Windowes of the House, on that Side, which the *Garden* stands, they be but Toyes: You may see as good Sights, many times, in Tarts. The *Garden* is best to be Square; Incompassed, on all the Foure Sides, with a *Stately Arched Hedge*. The *Arches* to be upon *Pillars*, of Carpenters Worke, of some Ten Foot high, and Six Foot broad: And the *Spaces* between, of the same Dimension, with the *Breadth* of the *Arch*. Over the *Arches*, let there bee an *Entire Hedge*, of some Foure Foot High, framed also upon Carpenters Worke: And upon the *Upper Hedge*, over every *Arch*, a little *Turret*, with a *Belly*, enough to receive a *Cage* of *Birds*: And over every *Space*, betweene the *Arches*, some other little *Figure*, with Broad Plates of *Round Coloured Glasse*, gilt, for the *Sunne*, to Play upon. But this *Hedge* I entend to be, raised upon a *Bancke*, not Steepe, but gently Slope, of some Six Foot, set all with *Flowers*. Also I understand, that this *Square* of the *Garden*, should not be the whole Breadth of the Ground, but to leave, on either Side, Ground enough, for diversity of *Side Alleys*: Unto which, the Two *Covert Alleys* of the *Greene*, may deliver you. But there must be, no *Alleys* with *Hedges*, at either *End*, of this great *Inclosure*: Not at the *Hither End*, for letting your Prospect upon this Faire Hedge from the *Greene*; Nor at the

Further End, for letting your Prospect from the Hedge, through the Arches, upon the *Heath*.

For the Ordering of the Ground, within the *Great Hedge*, I leave it to Variety of Device; Advising neverthelesse, that whatsoever forme you cast it into, first it be not too Busie, or full of Worke. Wherein I, for my part, doe not like *Images Cut out* in *Juniper*, or other *Garden stuffe*: They be for Children. *Little low Hedges*, Round, like Welts, with some Pretty *Pyramides*, I like well: And in some Places, *Faire Columnes* upon Frames of Carpenters Worke. I would also, have the *Alleys*, Spacious and Faire. You may have *Closer Alleys* upon the *Side Grounds*, but none in the *Maine Garden*. I wish also, in the very Middle, a *Faire Mount*, with three Ascents, and Alleys, enough for foure to walke a breast, Which I would have to be Perfect Circles, without any Bulwarkes, or Imbosments; And the *Whole Mount*, to be Thirty Foot high; And some fine *Banquetting House*, with some *Chimneys* neatly cast, and without too much *Glasse*.

For *Fountaines*, they are a great Beauty, and *Refreshment*; But *Pooles* marre all, and make the *Garden* unwholsome, and full of Flies, and Frogs. *Fountaines* I intend to be of two Natures: The One, that *Sprinckleth* or *Spouteth Water*; The Other a *Faire Receipt* of *Water*, of some Thirty or Forty Foot Square, but without Fish, or Slime, or Mud. For the first, the *Ornaments* of *Images Gilt*, or of *Marble*, which are in use, doe well: But the maine Matter is, so to Convey the Water, as it never Stay, either in the Bowles, or in the Cesterne; That the Water be never by Rest *Discoloured*, *Greene*, or *Red*, or the like; Or gather any *Mossinesse* or *Putrefaction*. Besides that, it is to be cleansed every day by the Hand. Also some *Steps* up to it, and some *Fine Pavement* about it, doth well. As for the other Kinde of *Fountaine*, which we may call a *Bathing Poole*, it may admit much Curiosity, and Beauty; wherewith we will not trouble our selves: As, that the Bottome be finely paved, And with Images: The sides likewise; And withall Embellished with Coloured Glasse, and such Things of Lustre; Encompassed also, with fine Railes of Low Statua's. But the Maine Point is the same, which we mentioned, in the former Kinde of *Fountaine*; which is, that the *Water* be in *Perpetuall Motion*, Fed by a Water higher then the *Poole*, and Delivered into it by

faire Spouts, and then discharged away under Ground, by some Equal-
itie of Bores, that it stay little. And for fine Devices, of Arching Water
without Spilling, and Making it rise in severall Formes, (of Feathers,
Drinking Glasses, Canopies, and the like,) they be pretty things to
looke on, but Nothing to Health and Sweetnesse.

For the *Heath*, which was the Third Part of our Plot, I wish it to be
framed, as much as may be, to a *Natural wildnesse*. Trees I would have
none in it; But some *Thickets*, made onely of *Sweet-Briar*, and honny-
suckle, and some *Wilde Vine* amongst; And the Ground set with
Violets, Strawberries, and *Prime-Roses*. For these are Sweet, and prosper in
the Shade. And these to be in the *Heath*, here and there, not in any
Order. I like also little *Heaps*, in the Nature of *Mole-hils*, (such as are in
Wilde Heaths) to be set, some with Wilde Thyme; Some with Pincks;
Some with Germander, that gives a good Flower to the Eye; Some
with Periwinckle; Some with Violets; Some with Strawberries;
Some with Couslips; Some with Daisies; Some with Red-Roses; Some
with Lilium Convallium; Some with Sweet-Williams Red; Some with
Beares-Foot; And the like Low Flowers, being withal Sweet, and Sight-
ly. Part of which *Heapes*, to be with *Standards*, of little *Bushes*, prickt
upon their Top, and Part without. The *Standards* to be Roses; Juniper;
Holly; Beare-berries (but here and there, because of the Smell of their
Blossome;) Red Currans; Goose-berries; Rose-Mary; Bayes; Sweet-
Briar; and such like. But these *Standards*, to be kept with Cutting, that
they grow not out of Course.

For the *Side Grounds*, you are to fill them with *Varietie* of *Alleys*,
Private, to give a full Shade; Some of them, wheresoever the Sun be.
You are to frame some of them likewise for Shelter, that when the
Wind blows Sharpe, you may walke, as in a Gallery. And those Alleys
must be likewise hedged, at both Ends, to keepe out the Wind; And
these *Closer Alleys*, must bee ever finely Gravelled, and no Grasse,
because of Going wet. In many of these *Alleys* likewise, you are to set
Fruit-Trees of all Sorts; As well upon the Walles, as in Ranges. And this
would be generally observed, that the *Borders*, wherin you plant your
Fruit-Trees, be Faire and Large, and Low, and not Steepe; And Set with
Fine Flowers, but thin and sparingly, lest they Deceive the Trees. At the

End of both the *Side Grounds*, I would have a *Mount* of some Pretty Height, leaving the Wall of the Enclosure Brest high, to looke abroad into the Fields.

For the *Maine Garden*, I doe not Deny, but there should be some Faire *Alleys*, ranged on both Sides, with *Fruit Trees*; And some Pretty *Tufts* of *Fruit Trees*, And *Arbours* with *Seats*, set in some Decent Order; But these to be, by no Meanes, set too thicke; But to leave the *Maine Garden*, so as it be not close, but the Aire Open and Free. For as for *Shade*, I would have you rest, upon the *Alleys* of the *Side Grounds*, there to walke, if you be Disposed, in the Heat of the Yeare, or day; But to make Account, that the *Maine Garden*, is for the more Temperate Parts of the yeare, And in the Heat of Summer, for the Morning, and the Evening, or Over-cast Dayes.

For *Aviaries*, I like them not, except they be of that Largenesse, as they may be *Turffed*, and have *Living Plants*, and *Bushes*, set in them; That the *Birds* may have more Scope, and Naturall Neastling, and that no *Foulenesse* appeare, in the *Floare* of the *Aviary*. So I have made a Platforme of a *Princely Garden*, Partly by Precept, Partly by Drawing, not a Modell, but some generall Lines of it; And in this I have spared for no Cost. But it is Nothing, for *Great Princes*, that for the most Part, taking Advice with Workmen, with no Lesse Cost, set their Things together; And sometimes adde *Statua's*, and such Things, for State, and Magnificence, but nothing to the true Pleasure of a *Garden*.

SIR FRANCIS BACON, 1625

If the Gentlemen of *England* had formerly been better advised in the laying out their Gardens, we might by this Time been at least equal (if not far superior) to any Abroad.

For as we abound in good Soil, fine Grass, and Gravel, which in many Places Abroad is not to be found, and the best of all Sorts of Trees; it therefore appears, that nothing has been wanting but a noble Idea of the Disposition of a Garden. I could instance divers Places in *England*, where Noblemen and Gentlemens Seats are very finely situated, but wretchedly executed, not only in respect to disproportion'd Walks, Trees planted in improper Soils, no Regard had to fine Views, &c. but with that abominable Mathematical Regularity and Stiffness, that nothing that's bad could equal them.

Now these unpleasant forbidding Sort of Gardens, owe their Deformity to the insipid Taste or Interest of some of our Theorical Engineers, who, in their aspiring Garrets, cultivate all the several Species of Plants, as well as frame Designs for Situations they never saw: Or to some Nursery-Man, who, for his own Interest, advises the Gentleman to such Forms and Trees as will make the greatest Draught out of his Nursery, without Regard to any Thing more: And oftentimes to a Cox-comb, who takes upon himself to be an excellent Draughtsman, as well as an incomparable Gardener; of which there has been, and are still, too many in *England*, which is witness'd by every unfortunate Garden wherein they come. Now as the Beauty of Gardens in general depends upon an elegant Disposition of all their Parts, which cannot be determined without a perfect Knowledge of its several Ascendings, Descendings, Views, &c. how is it possible that any Person can make a good Design for any Garden, whose Situation they never saw?

To draw a beautiful regular Draught, is not to the Purpose; for altho' it makes a handsome Figure on the Paper, yet it has a quite different Effect when executed on the Ground: Nor is there any Thing more ridiculous, and forbidding, than a Garden which is regular; which, instead of entertaining the Eye with fresh Objects, after you have seen a quarter Part, you only see the very same Part repeated again, without any Variety.

And what still greatly adds to this wretched Method, is, that to execute these stiff regular Designs, they destroy many a noble Oak, and in its Place plant, perhaps, a clumsey-bred Yew, Holley, &c. which, with me, is a Crime of so high a Nature, as not to be pardon'd.

There is nothing adds so much to the Pleasure of a Garden, as those great Beauties of Nature, *Hills* and *Valleys*, which, by our *regular Coxcombs*, have ever been destroyed, and at a very great Expence also in Levelling.

For, to their great Misfortune, they always deviate from Nature, instead of imitating it.

There are many other Absurdities I could mention, which those *wretched Creatures* have, and are daily guilty of: But as the preceding are sufficient to arm worthy Gentlemen against such Mortals, I shall at present forbear . . .

BATTY LANGLEY, 1728

If we consider the Works of *Nature* and *Art*, as they are qualified to entertain the Imagination, we shall find the last very defective, in Comparison of the former; for though they may sometimes appear as Beautiful or Strange, they can have nothing in them of that Vastness and Immensity, which afford so great an Entertainment to the Mind of the Beholder. The one may be as Polite and Delicate as the other, but can never shew herself so August and Magnificent in the Design. There is something more bold and masterly in the rough careless Strokes of Nature, than in the nice Touches and Embellishments of Art. The Beauties of the most stately Garden or Palace lie in a narrow Compass, the Imagination immediately runs them over, and requires something else to gratify her; but, in the wide Fields of Nature, the Sight wanders up and down without Confinement, and is fed with an infinite variety of Images, without any certain Stint or Number. For this Reason we always find the Poet in Love with a Country-Life, where Nature appears in the greatest Perfection, and furnishes out all those Scenes that are most apt to delight the Imagination . . .

But tho' there are several of these wild Scenes, that are more delightful than any artificial Shows; yet we find the Works of Nature still more pleasant, the more they resemble those of Art: For in this case our Pleasure rises from a double Principle; from the Agreeableness of the Objects to the Eye, and from their Similitude to other Objects: We are pleased as well with comparing their Beauties, as with surveying them, and can represent them to our Minds, either as Copies or Originals. Hence it is that we take Delight in a Prospect which is well laid out, and diversified with Fields and Meadows, Woods and Rivers; in those accidental Landskips of Trees, Clouds and Cities, that are sometimes found in the Veins of Marble; in the curious Fret-work of Rocks and Grottos; and, in a Word, in any thing that hath such a Variety or Regularity as may seem the Effect of Design in what we call the Works of Chance.

If the Products of Nature rise in Value, according as they more or less resemble those of Art, we may be sure that artificial Works receive a greater Advantage from their Resemblance of such as are natural; because here the Similitude is not only pleasant, but the Pattern more perfect. The prettiest Landskip I ever saw, was one drawn on the Walls of a dark Room, which stood opposite on one side to a navigable River, and on the other to a Park. The Experiment is very common in Opticks. Here you might discover the Waves and Fluctuations of the Water in strong and proper Colours, with the Picture of a Ship entring at one end, and sailing by Degrees through the whole Piece. On another there appeared the Green Shadows of Trees, waving to and fro with the Wind, and Herds of Deer among them in Miniature, leaping about upon the Wall. I must confess, the Novelty of such a Sight may be one occasion of its Pleasantness to the Imagination, but certainly the chief Reason is its near Resemblance to Nature, as it does not only, like other Pictures, give the Colour and Figure, but the Motion of the Things it represents.

We have before observed, that there is generally in Nature something more Grand and August, than what we meet with in the Curiosities of Art. When, therefore, we see this imitated in any measure, it gives us a nobler and more exalted kind of Pleasure than what we receive from the nicer and more accurate Productions of Art. On this Account our *English* Gardens are not so entertaining to the Fancy as those in *France* and *Italy*, where we see a large Extent of Ground covered over with an agreeable Mixture of Garden and Forest, which represent every where an artificial Rudeness, much more charming than that Neatness and Elegancy which we meet with in those of our own Country. It might, indeed, be of ill Consequence to the Publick, as well as unprofitable to private Persons, to alienate so much Ground from Pasturage, and the Plow, in many Parts of a Country that is so well peopled, and cultivated to a far greater Advantage. But why may not a whole Estate be thrown into a kind of Garden by frequent Plantations, that may turn as much to the Profit, as the Pleasure of the Owner? A Marsh overgrown with Willows, or a Mountain shaded with Oaks, are not only more beautiful, but more beneficial, than when they lie bare and unadorned. Fields of Corn make a pleasant Prospect, and if the Walks were a little taken care of that lie between them, if the natural Embroidery of the Meadows were helpt and improved by some small Additions of Art, and the several Rows of Hedges set off by Trees and Flowers, that the soil was capable of receiving, a Man might make a pretty Landskip of his own Possessions.

Writers, who have given us an Account of *China*, tell us the Inhabitants of that Country laugh at the Plantations of our *Europeans*, which are laid out by the Rule and Line; because, they say, any one may place Trees in equal Rows and uniform Figures. They chuse rather to shew a Genius in Works of this Nature, and therefore always conceal the Art by which they direct themselves. They have a Word it seems in their Language, by which they express the particular Beauty of a Plantation that thus strikes the Imagination at first Sight, without discovering what it is that has so agreeable an Effect. Our *British* Gardeners, on the contrary, instead of humouring Nature, love to deviate from it as much as possible. Our Trees rise in Cones, Globes, and Pyramids. We see the Marks of the Scissars upon every Plant and Bush. I do not know whether I am singular in my Opinion, but, for my own part, I would rather look upon a Tree in all its Luxuriancy and Diffusion of Boughs and Branches, than when it is thus cut and trimmed into a

Mathematical Figure; and cannot but fancy that an Orchard in Flower looks infinitely more delightful than all the little Labyrinths of the most finished Parterre. But as our great Modellers of Gardens have their Magazines of Plants to dispose of, it is very natural for them to tear up all the Beautiful Plantations of Fruit Trees, and contrive a Plan that may most turn to their own Profit, in taking off their Evergreens, and the like Moveable Plants, with which their Shops are plentifully stocked.

JOSEPH ADDISON, 1712

... simplicity and grandeur of taste in gardening, which has produced many fine plantations in this kind, is at present suffering with that of all other things; the caracatura and minute are again prevailing in too many places.

The citizen who visits his rural retirement close to the road, thronged with coaches, carts, waggons, chaises, and all kinds of carriages, which differs from London only in this, that in winter it rains smoke in the city, and in summer dust in the country, must have his plantation of an acre diversified with all that is to be found in the most extensive garden of some thousand acres; here must be temples to every goddess as well as Cloacina, woods, waters, lawns, and statues, which being thus contrived to contain so many things, is in fact, nothing at all, and that which might be something by being but one, is entirely lost by being intended to be so many; one wonders how so many things can be crammed into so small a place, as we do at the whole furniture of a room in a cherry stone; it is a scene for fairies.

JOHN SHEBBEARE, 1755

Although music, poetry, and painting, sister fine arts, have in all enlightened countries sooner arrived at perfection than Landscape Gardening, yet the latter offers to the cultivated mind in its more perfect examples, in a considerable degree a union of all these sources of enjoyment; a species of *harmony*, in a pleasing combination of the most fascinating materials of beauty in natural scenery: *poetic* expression in the babbling brook, the picturesque wood, or the peaceful sunlit turf: and the lovely effects of landscape *painting*, realized in the rich, varied, and skillfully arranged whole.

The object of this charming art, is to create in the grounds of a country residence a kind of polished scenery, producing a delightful effect, either by a species of studied and elegant *design*, in symmetrical or regular plantations: or by a combination of beautiful or picturesque forms, such as we behold in the most captivating passages of general nature.

The *practice* of Landscape Gardening has grown out of that love of country life, and the desire to render our own property attractive, which naturally exists to a greater or lesser degree in the minds of all men. In the case of large landed estates, the capabilities of Landscape Gardening may be displayed to their full extent, as from five to five hundred acres may be devoted to a park or pleasure-grounds. But the principles of the art may be applied, and its beauties realized to a certain degree, in the space of half an acre of ground – wherever grass will grow, and trees thrive luxuriantly.

Two distinct modes of the art widely differing in themselves, have divided, for some time, the admiration of the world. One is the Ancient, formal or Geometric Style: the other the Modern, Natural, or Irregular Style. The first, characterized by regular forms and right lines, the last by varied forms and flowing lines.

A recurrence to the history of Gardening as well as to the history of the fine arts, will afford abundant proof that in the first stage, or infancy of these arts, while the perception of their ultimate capabilities is yet crude and imperfect, mankind has in every instance been completely satisfied with the mere exhibition of *design* or *art*. Thus in Sculpture, the first statues were only attempts to imitate rudely the *form* of a human figure, or in painting, to represent that of a tree: the skill of the artist in effecting an imitation successfully, was sufficient to excite the astonishment and admiration of those who had not yet made such advances as to enable them to appreciate the superior beauty of *expression*.

In laying out gardens, the practice from all antiquity (until in late times the superiority of natural beauty was discovered) has been to display the skill of the designer in arranging all the materials of nature, in artificial, regular, or symmetrical forms. Walks and roads straight, beds square and round, trees smoothly clipped and shorn into different figures, these were the predominant characteristics of the Ancient or Geometric Style. That person who possessed in his grounds a luxuriant and graceful elm, its branches elegantly sweeping the earth, and forming a varied outline against the sky, saw no more than nature everywhere afforded: but he, whose garden exhibited a cypress or a yew cut by the shears into a four-sided pyramid of verdure, had at least achieved something which nature has not been able to do, and commanded a sort of respect for the excellence or novelty of his art. This taste rendered more or less elegant, continued throughout all Europe until about the year 1700. The lavish expenditure in the royal and princely gardens of the courts of Europe, in the decoration and embellishment of their gardens, gave a new impulse as well as a sublime grandeur, to the art. The finest example of this style is perhaps that of Versailles, the garden of the extravagant Louis XIV, and the most distinguished artist its designer, Le Nôtre. Its water works, *jets-d'eau*, etc., alone, are stated to have been played off seven or eight times a year at an expense of more than two thousand dollars per hour. Sculpture of every description, mural and verdant, was scattered in profusion through the superb gardens of this period; statues and busts of celebrated heroes and statesmen, fountains of all descriptions, urns and vases almost without number: and the whole, especially on so grand a scale, had a most imposing and magnificent effect.

Any person who will analyze the kind of beauty aimed at in the ancient style, will we think, at once perceive its characteristics to be *uniformity* and the display of *symmetric art*. Almost any one may succeed in laying out and planting a garden in right lines, and may give it an air of stateliness and grandeur, by costly decorations; and even now, there are perhaps thousands who would express greater delight in walking through such a garden, than in surveying one where the finest natural beauties are combined. The reason of this is indeed sufficiently obvious.

Every one, though possessed of the least possible portion of taste, readily appreciates the cost and labour incurred in the first case, and bestows his admiration accordingly; but we must infer the presence of a cultivated and refined mind, to realize and enjoy the more exquisite beauty of natural forms.

As however cultivation progressed in Europe, the taste for this style began to be weakened by several causes. In the first place, a large portion of the lands coming under the plough, fine natural woods were gradually cut off, and wild landscape beauties, once so common as to be unheeded, became sufficiently rare to be more prized and admired. The increased admiration of landscape painting, poetry, and other fine arts, by imbuing many minds with a love of beautiful and picturesque nature, also tended to create a change in the taste. Gradually, men of refined sensibilities perceived that besides mere beauty of FORM, natural objects have another and much higher kind of beauty – namely, the beauty of EXPRESSION.

With the recognition of this principle commenced a new era in Ornamental Gardening. The defects of the Geometric School, were freely pointed out and discussed, by writers of cultivated sensibility and taste, and an entire revolution suddenly took place in the public mind. With a higher perception of the capacities of Landscape Gardening, gradually grew up another class of artists, who, laying aside the prejudice which allowed men to see beauty in Gardens, only through the manifestation of design, derived from the study of nature, new elements to interest the mind as well as elevate the art. One of these, looking around him for materials, observes the spirit and expression of natural objects, the varied forms of ground and water, and the character of trees individually and in composition. He perceives that there is an expression of dignity and majesty in an old oak, of grace-fulness and luxuriance in a fine sweeping elm, or of the spirited and picturesque in the larch, which confer or create a character in scenes in which they are happily introduced: and, laying down the shears of the old gardeners, he feels that there is a grace and beauty in their free and unshorn luxuriance, infinitely above that of the tree, clipped according to the rules of a formal art. Undulating surfaces of ground have an expression superior to the tame level; and there is a more delightful variety in a walk of half a mile in length, which winds naturally here and there, over a diversified surface, bordered occasionally with luxuri-ant groups of trees, open spaces of fine lawn, and dense thickets of shrubbery, or underwood, than in a straight level avenue over the same distance, whose sides present but one continuous line of trees seen at the same moment, and presenting but one single and monotonous view. Losing by degrees his reverence for avowed and uniform art, he learns to appreciate those flowing, smooth, and continuous lines, which characterize objects the most graceful and delicate around us; in short that, instead of endeavouring to distort Nature, we should rather strive to heighten her beauties and remove all her defects.

ANDREW JACKSON DOWNING, 1841

The ordinary modern flower-bed is ugly in form and monotonous in colour, and it seems to be thought necessary to border it with the ugly lobelia, regardless of the colours of the flower-bed itself. All the fancy has gone out of it, and little or no attempt is made to lay out the beds on any consecutive scheme. Contrast this with the beds of the old gardens of New College [Oxford], now destroyed. In front of the entrance gateway there was a broad path about eighteen feet wide, with cross paths subdividing the garden into four square plots.

On the right-hand plot as you entered was worked, probably in rosemary, hyssop, or thyme, the arms of New College and the motto
'manners makyth man' and the date. In the next plot was a curious
device in flowers. On the left hand was planted the royal arms and the
date 1628; and the plot beyond was laid out as an enormous sundial,
the hours probably shown in box or rosemary on an oval of sand, with
an upright dial formed of wood in the centre . . . Such a garden, if it
had been preserved, would have been beyond all price to us now. The
ridicule with which such work is dismissed, the abuse lavished on it as
artificial, is beside the mark. It is just this very artifice, this individual-
ity, this human interest, that gives to the old formal garden its undying
charm – the feeling that once there was a man or a woman who cared
about the garden enough to have it laid out in one way more than
another, and that they and many generations since have taken pleasure
in its beauty and the fancy of its parterres. Perhaps, when any tradition
of art is formed among us again, there will return this pleasure and
delight in those old ways which are the better.

SIR REGINALD BLOMFIELD, 1892

I am friendly to the fashion of laying out flower gardens on grass
lawns, surrounded with a shrubbery of the choicest species, and at
proper distances, clusters of Hollyhocks, Dahlias, Delphiniums or
Bee-Larkspurs, Heleanthemums or perennial Sun Flowers, Rud-
beckias, Solidagos or Golden Rods, Starry Asters or Michaelmas
Daisies, &c.; those may be placed promiscuously in the fore-ground
of the shrubbery. Walks should judiciously intersect the plantings, in
such manner as to lead to the most advantageous points for viewing
the flower garden and pleasure grounds.

ALEXANDER FORBES, 1820

The formal treatment of gardens ought, perhaps, to be called the architectural treatment of gardens, for it consists in the extension of the principles of design which govern the house to the grounds which surround it. Architects are often abused for ignoring the surroundings of their buildings in towns, and under conditions which make it impossible for them to do otherwise; but if the reproach has force, and it certainly has, it applies with greater justice to those who control both the house and its surroundings, and yet deliberately set the two at variance. The object of formal gardening is to bring the two into harmony, to make the house grow out of its surroundings, and to prevent its being an excrescence on the face of nature. The building cannot resemble anything in nature, unless you are content with a mud-hut and cover it with grass. Architecture in any shape has certain definite characteristics which it cannot get rid of; but, on the other hand, you can lay out the grounds, and alter the levels, and plant hedges and trees exactly as you please; in a word, you can so control and modify the grounds as to bring nature into harmony with the house, if you cannot bring the house into harmony with nature. The harmony arrived at is not any trick of imitation, but an affair of a dominant idea which stamps its impress on house and grounds alike.

SIR REGINALD BLOMFIELD, 1892

I am frankly and absolutely for a formal garden. This may turn you away from me, but I hope not. Once and for all I declare against the thing called 'landscape-gardening', and cleave to classic precedents. Note the high tone I take in this matter. With a house like mine there is really some excuse for seeking to ignore it, and developing a garden that shall be independent of architecture so dreadful; but no, I will be just; my garden shall shame my house by its correct proportions and proper adherence to what a garden ought to be. Not that this garden is classic – far from that; I wish it was. But it is a garden, no mere feeble deception. It is a small piece of ground enclosed by walls; and, concerning those walls, you are in no doubt for one moment. There is not the least attempt to imitate natural scenery. There are no winding walks, no boskages, no sylvan dells, no grottoes stuck with stones and stalactites. My garden is simply an artificial, but none the less beautiful, arrangement of all the best plants that I can contrive to collect.

Consider the word 'garden'. It develops by evolution from the Anglo-Saxon 'geard' and the Middle English 'garth'. It means 'a yard'. It has rather less than nothing to do with wild nature, or any other sort of nature. It is a highly artificial contrivance within hard and fast boundaries. We speak of a zoological garden, a garden of pleasure, a garden of vegetables. To talk of a 'natural' or a 'wild' garden, is a contradiction in terms. You might as well talk of a natural 'zoo', and do away with the bars, and arrange bamboo brakes for the tigers, mountain-tops for the eagles, and an iceberg for the polar bears.

EDEN PHILLPOTTS, 1906

Why should we imitate wild nature? The garden is a product of civilization. Why any more make of our gardens imitation of wild nature, than paint our children with woad, and make them run about naked in an effort to imitate nature unadorned? The very charm of a garden is that it is taken out of savagery, trimmed, clothed and disciplined.

S. BARING-GOULD, 1890

What I have said of the best forms of gardens, is meant only of such as are in some sort regular; for there may be other forms wholly irregular, that may, for ought I know, have more beauty than any of the others; but they must owe it to some extraordinary dispositions of nature in the seat, or some great race of fancy or judgment in the contrivance, which may produce many disagreeing parts into some figure, which shall yet upon the whole, be very agreeable. Something of this I have seen in some places, but heard more of it from others, who have lived much among the Chinese; a people, whose way of thinking seems to lie as wide of ours in Europe, as their country does. Among us, the beauty of building and planting is placed chiefly in some certain proportions, symmetries, or uniformities; our walks and our trees ranged so, as to answer one another, and at exact distances. The Chinese scorn this way of planting, and say a boy that can tell an hundred, may plant walks of trees in straight lines, and over against one another, and to what length and extent he pleases. But their greatest reach of imagination, is employed in contriving figures, where the beauty shall be great, and strike the eye, but without any order or disposition of parts, that shall be commonly or easily observed. And though we have hardly any notion of this sort of beauty, yet they have a particular word to express it; and where they find it hit their eye at first sight, they say the *Sharawadgi* is fine or is admirable, or any such expression of esteem. And whoever observes the work upon the best Indian gowns, or the painting upon their best screens or purcellans, will find their beauty is all of this kind, [that is] without order. But I should hardly advise any of these attempts in the figure of gardens among us; they are adventures of too hard achievement for any common hands; and though there may be more honour if they succeed well, yet there is more dishonour if they fail, and 'tis twenty to one they will; whereas in regular figures, 'tis hard to make any great and remarkable faults.

SIR WILLIAM TEMPLE, 1685 (pub. 1692)

To improve the scenery of a country, and to display its native beauties with advantage, is an ART which originated in England, and has therefore been called *English Gardening*; yet as this expression is not sufficiently appropriate, especially since Gardening, in its more confined sense of *Horticulture*, has been likewise brought to the greatest perfection in this country, I have adopted the term *Landscape Gardening*, as most proper, because the art can only be advanced and perfected by the united powers of the *landscape painter* and the *practical gardener*. The former must conceive a plan, which the latter may be able to execute; for though the painter may represent a beautiful landscape on his canvass, and even surpass Nature by the combination of her choicest materials, yet the luxuriant imagination of the *painter* must be subjected to the *gardener's* practical knowledge in planting, digging, and moving earth; that the simplest and readiest means of accomplishing each design may be suggested; since it is not by vast labour, or great expense, that Nature is generally to be improved; on the contrary,

> Ce noble emploi demande un artiste qui pense,
> Prodigue de génie, mais non pas de dépense.

HUMPHRY REPTON, 1795

Among my reasons for thinking wild gardening worth practising by all who wish our gardens to be more artistic and delightful are the following: –

First, because hundreds of the finest hardy flowers will thrive much better in rough places than ever they did in the old-fashioned border. Even small plants, like the ivy-leaved Cyclamen, a beautiful plant that we rarely find in perfection in gardens, I have seen perfectly naturalized and spread all over the mossy surface of a thin wood.

Secondly, because they will look infinitely better than they ever did in formal beds, in consequence of fine-leaved plant, fern, and flower, and climber, grass and trailing shrub, relieving each other in delightful ways. Many arrangements will prove far more beautiful than any aspect of the old mixed border, or the ordinary type of modern flower-garden.

Thirdly, because no disagreeable effects result from decay. The raggedness of the old mixed border after the first flush of spring and early summer bloom had passed was intolerable to many, with its bundles of decayed stems tied to sticks. When Lilies are sparsely dotted through masses of shrubs, their flowers are admired more than if they were in isolated showy masses; when they pass out of bloom they are unnoticed amidst the vegetation, and not eyesores, as when in rigid unrelieved tufts in borders, &c. In a semi-wild state the beauty of a fine plant will show when at its height; and when out of bloom it will be followed by other kinds of beauty.

Fourthly, because it will enable us to grow many plants that have never yet obtained a place in our 'trim gardens'. I mean plants which, not so showy as many grown in gardens, are never seen therein. The flowers of many of these are of great beauty, especially when seen in numbers. A tuft of one of these in a border may not be thought worthy of its place, while in some wild glade, as a little colony, grouped naturally, its effect may be exquisite. There are many plants too that, grown in gardens, are no great aid to them – like the Golden Rods, and other plants of the great order Compositæ, which merely overrun the choicer and more beautiful border-flowers when planted amongst them. These coarse plants would be quite at home in copses and woody places, where their blossoms might be seen or gathered in due season, and their vigorous vegetation form a covert welcome to the game-preserver. To these two groups might be added plants like the winter Heliotrope, and many others which, while not without use in the garden, are apt to become a nuisance there. For instance, the Great Japanese Knotworts (Polygonum) are certainly better planted outside of the flower-garden.

Fifthly, because we may in this way settle the question of the spring flower-garden. Many parts of every country garden, and many suburban ones, may be made alive with spring flowers, without interfering at least with the flower-beds near the house. The blue stars of the Apennine Anemone will be enjoyed better when the plant is taking care of itself, than in any conceivable formal arrangement. It is but one of hundreds of sweet spring flowers that will succeed perfectly in our fields, lawns, and woods. And so we may cease the dreadful practice of tearing up the flower-beds and leaving them like new-dug graves twice a year.

Sixthly, because there can be few more agreeable phases of communion with Nature than naturalizing the natives of countries in which we are infinitely more interested than in those of which greenhouse or stove plants are native. From the Roman ruin – home of many flowers, the mountains and prairies of the New World, the woods and meadows of all the great mountains of Europe; from Greece and Italy and Spain, from the hills of Asia Minor; from the alpine regions of the great continents – in a word, from almost every interesting region the traveller may bring seeds or plants, and establish near his home living souvenirs of the various countries he has visited. If anything we may bring may not seem good enough for the garden autocrat of the day, it may be easy to find a home for it in wood or hedgerow; I am fond of putting the wild species of Clematis and other exotic climbers and flowers in newly-formed hedgebanks.

Moreover, the great merit of permanence belongs to this delightful phase of gardening. Select a rough slope, and embellish it with groups of the hardiest climbing plants, – say the Mountain Clematis from Nepal, the sweet C. Flammula from Southern Europe, 'Virginian creepers', various hardy vines, Jasmines, Honeysuckles, and wild Roses and briers. Arranged with some judgment at first, such a colony might be left to take care of itself; time would but add to its attractions.

Some have mistaken the idea of the wild garden as a plan to get rid of all formality near the house; whereas it will restore to its true use the flower-garden, now subjected to two tearings up a year – i.e. in spring and autumn; as may be seen in nearly all public and private gardens, in France as well as in England – new patterns every autumn and every spring – no rest or peace anywhere. In the beautiful summer of 1893, the flower-beds in the public gardens of Paris were quite bare of all flowers in June, before the wretched winter-nursed flowers had been set out in their patterns. If such things must be done in the name of flower-gardening, it were many times better to carry them out in a place apart, rather than expose the foreground of a beautiful house or landscape to such disfigurement. Spring flowers are easily grown in multitudes away from the house, and, therefore, for their sakes the system of digging up the flower-beds twice a year need not be carried out. Wild gardening should go hand in hand with the thorough culti-vation of the essential beds of the flower-garden around the house, and to their being filled with plants quite different from those we entrust to the crowded chances of turf or hedgerow: – to rare or tender plants or choice garden flowers like the Tea Rose and Carnation – plants which often depend for their beauty on their double states, and for which rich soil and care and often protection are essential.

WILLIAM ROBINSON, 1870 (5th edn, 1895)

I live near one of the most beautiful so-called wild gardens in England, but it requires endless care, and is always extending in all directions in search of fresh soil. What is possible is to have the appearance of a wild garden in consequence of the most judicious planting, with consummate knowledge and experience of the plants that will do well in the soil if they are just a little assisted at the time of planting. I saw the other day the most lovely Surrey garden I know, though it is without any peculiar natural advantages from the lie of the land – a flat piece of ground on the top of a hill, a copse wood of Spanish Chestnut, Birch, Holly, and Fir. Even in the original thinning of the wood the idea had been formulated that certain plants and trees had better be kept together as they grew, and broad open spaces had been cut, broken up with groups of Holly for protection. The paths were laid with that short turf that grows on Surrey commons, and only wants mowing three or four times a year. The planting had been done with the greatest skill, almost imperceptibly getting more and more cared for and refined as it got nearer the house. Here I saw, among many other things, the finest specimens of the smaller Magnolias, *Stellata* and *Conspicua*. This surprised me, as I thought they required heavy soil. The ground had been thoroughly well made, and they were well away from any trees that could rob them; but in the lightest, dryest soil they were far finer plants than the specimen plants in the grass lawns at Kew. This whole garden was such a beautiful contrast from the usual planning and clearing-away of all the natural advantages that generally surround a place which is being built or altered. The land, as a rule, is dug over and made flat, and planted in the usual horrible shrubbery style. I have seen such wonderful natural advantages thrown away, a copse laid low to extend a lawn, a lovely spring, which could have been turned into a miniature river, made into a circular pond, with Laurels, Rhododendrons, and other shrubs dotted about, and twisted gravel paths made round it. Another lovely natural pond I knew, into which the rains drained, though nearly at the top of a hill, where water was precious and scarce. Now it is cemented all round with hard, cold cement, on which nothing can grow, and into which, in the wettest of weather, the water can no longer drain. The pond never fills, and nothing can grow around it. I know few things more depressing than an utter want of feeling for Nature's ways of playing the artist, as she does at every turn. I cannot understand anyone walking down a hilly road after rain without admiring the action of the water on its surface, with the beautiful curves and turns and sand islands that Nature leaves in playful imitation of her grandest efforts ... It has long been said, 'God sends the food, and the Devil sends the cook.' I am sure the same might be said of the owners, the nurserymen, and the landscape gardeners, who most carefully, as a rule, throw away every single natural advantage of the piece of ground they are laying out, and believe they are 'improving'! What would give me the greatest pleasure would be to have the laying-out of a little place on the side of a hill with a fine view to the south and west, and the land sloping away and gently terraced till it reached the plain at the bottom.

MRS C. W. EARLE, 1897

Well knowing that every situation has its facilities and its difficulties, I have never considered how many acres I was called upon to improve, but how much I could improve the subject before me ...

Some of the places ... are subjects which I have visited only once; others, from the death of the proprietors, the change of property, the difference of opinions, or a variety of other causes, may not, perhaps, have been finished according to my suggestions. It would be endless to point out the circumstances in each place where my plans have been partially adopted or partially rejected. To claim as my own, and to arrogate to myself all that I approve at each place, would be doing injustice to the taste of the several proprietors who may have suggested improvements. On the other hand, I should be sorry, that to *my taste* should be attributed all the absurdities which fashion, or custom, or whim, may have occasionally introduced in some of these places. I can only advise, I do not pretend to dictate, and, in many cases, must rather conform to what has been ill begun, than attempt to pull to pieces and remodel the whole Work . . .

To avoid the imputation of having fully *approved*, where I have found it necessary merely to *assent*, I shall here beg leave to subjoin my opinion negatively, as the only means of doing so without giving offence to those from whom I may differ; at the same time, with the humility of experience, I am conscious my opinion may, in some cases, be deemed wrong. The same motives which induce me to mention what I recommend, will also justify me in mentioning what I disapprove; a few observations, therefore, are subjoined to mark those errors, or absurdities in modern gardening and architecture, to which I have never willingly subscribed, and from which it will easily be ascertained how much of what is called the improvement of any place in the list, may properly be attributed to my advice. It is rather upon my opinions in writing, than on the partial and imperfect manner in which my plans have sometimes been executed, that I wish my fame to be established.

Objection No. 1.

There is no error more prevalent in modern gardening, or more frequently carried to excess, than taking away hedges to unite many small fields into one extensive and naked lawn, before plantations are made to give it the appearance of a park; and where ground is subdivided by sunk fences, imaginary freedom is dearly purchased at the expense of actual confinement.

No. 2.

The baldness and nakedness round the house is part of the same mistaken system, of concealing fences to gain extent. A palace, or even an elegant villa, in a grass field, appears to me incongruous; yet I have seldom had sufficient influence to correct this common error.

No. 3.

An approach which does not evidently lead to the house, or which does not take the shortest course, cannot be right. [This rule must be taken with certain limitations. The shortest road across a lawn to a house will seldom be found graceful, and often vulgar. A road bordered by trees in the form of an avenue, may be straight without being vulgar; and grandeur, not grace or elegance, is the expression expected to be produced.]*

No. 4.

A poor man's cottage, divided into what is called *a pair of lodges*, is a mistaken expedient to mark importance in the entrance to a Park.

No. 5.

The entrance gate should not be visible from the mansion, unless it opens into a court yard.

No. 6.

The plantation surrounding a place, called a *Belt*, I have never advised; nor have I ever willingly marked a drive, or walk, completely round the verge of a park, except in small villas, where a dry path round a person's own field is always more interesting to him than any other walk.

*Note by John Claudius Loudon in his 1840 edition.

No. 7.

Small plantations of trees, surrounded by a fence, are the best expedients to form groups, because trees planted singly seldom grow well; neglect of thinning and removing the fence, has produced that ugly deformity called a *Clump*.

No. 8.

Water on an eminence, or on the side of a hill, is among the most common errors of Mr ['Capability'] Brown's followers: in numerous instances I have been allowed to remove such pieces of water from the hills to the valleys; but in many my advice has not prevailed.

No. 9.

Deception may be allowable in imitating the works of NATURE; thus artificial rivers, lakes, and rock scenery, can only be great by deception, and the mind acquiesces in the fraud, after it is detected: but in works of ART every trick ought to be avoided. Sham churches, sham ruins, sham bridges, and everything which appears what it is not, disgusts when the trick is discovered.

No. 10.

In buildings of every kind the *character* should be strictly observed. No incongruous mixture can be justified. To add Grecian to Gothic, or Gothic to Grecian, is equally absurd; and a sharp pointed arch to a garden gate or a dairy window, however frequently it occurs, is not less offensive than Grecian architecture, in which the standard rules of relative proportion are neglected or violated.

The perfection of landscape gardening consists in the fullest attention to these principles, *Utility*, *Proportion*, and *Unity*, or harmony of parts to the whole.

HUMPHRY REPTON, 1840

Flowers in masses are mighty strong colour, and if not used with a great deal of caution are very destructive to pleasure in gardening. On the whole, I think the best and safest plan is to mix up your flowers, and rather eschew great masses of colour – in combination I mean. But there are some flowers (inventions of men, i.e. florists) which are bad colour altogether, and not to be used at all. Scarlet geraniums, for instance, or the yellow calceolaria, which indeed are not uncommonly grown together profusely, in order, I suppose, to show that even flowers can be thoroughly ugly.

Another thing also much too commonly seen is an aberration of the human mind, which otherwise I should have been ashamed to warn you of. It is technically called carpet-gardening. Need I explain it further? I had rather not, for when I think of it even when I am quite alone I blush with shame at the thought.

WILLIAM MORRIS, 1882

I see too many gardens spoilt because stone is rashly introduced into a countryside where brick is the universal building material and vice versa. In the same way new materials – concrete, synthetic stone, glass, mosaic, faience and plastics – all have their uses and possibilities, but only where they are not used in a context with traditional or historical associations. To use these new materials carelessly is to risk dating a garden, for without the most brilliant and yet restrained handling they will indicate 'fashion' rather than style and so will become fatally un-fashionable in twenty years. Only if the garden survives a half-century or longer, might they achieve 'quaintness' and have the dubious charm that the bark-houses, ferneries and felspar grottoes of a hundred or more years ago sometimes still exercise on the romantically-inclined spectator.

RUSSELL PAGE, 1962

A flower-garden should be an object detached and distinct from the general scenery of the place; and, whether large or small, whether varied or formal, it ought to be protected from hares and smaller animals by an inner fence: within this enclosure rare plants of every description should be encouraged, and a provision made of soil and aspect for every different class. Beds of bog-earth should be prepared for the American plants: the aquatic plants, some of which are peculiarly beautiful, should grow on the surface or near the edges of water. The numerous class of rock-plants should have beds of rugged stone provided for their reception, without the affectation of such stones being the natural production of the soil; but, above all, there should be poles or hoops for those kind of creeping plants which spontaneously form themselves into graceful festoons, when encouraged and supported by art.

Yet, with all these circumstances, the flower-garden, except where it is annexed to the house, should not be visible from the roads or general walks about the place. It may therefore be of a character totally different from the rest of the scenery, and its decorations should be as much those of art as of nature.

HUMPHRY REPTON, 1795

Everyone who can, now lives in the country, where he is bound to have a garden; and ... let no one suppose that the beauty of a garden depends on its acreage, or on the amount of money spent upon it. Nay, one would almost prefer a small garden plot, so as to ensure that ample justice shall be done to it. In a small garden there is less fear of dissipated effort, more chance of making friends with its inmates, more time to spare to heighten the beauty of its effects.

JOHN D. SEDDING, 1890

In the first place, you are if you may conveniently, to erect [the Garden of Pleasure] in such a place where it may yield most delight, in regard of its prospect from your House, or some chief Rooms thereof; and withal, if it may be pretty well defended from the injury of the sharpest winds; and in so doing, you may have in a manner a perpetual Spring, something or other continually in its beauty, either Flowers, or ever Greens, except in extream Frost and Snow, but even then there are many housed greens do shew forth their Beauties, but let every one do as their means, minds, or conveniences will permit.

The plot of ground being resolv'd upon, you are free to fence it in according as you desire, or can; only remember that if there need either bringing in, or carrying out of mould, *&c.* that you do it whilst you have the opportunity of a Cart-way, which is usually cheaper and speedier than Wheel-barrows, and then you may level it, and craft it into what form you think fit, or as the bigness of your ground will handsomely bear.

I have for the ease and delight of those that do affect such things, presented to view divers forms or plots for Gardens, amongst which it is possible you may find some that may near the matter fit most ordinary grounds, either great or small; and shall leave the ingenious Practitioner to the consideration and use of that he most affects.

LEONARD MEAGER, 1688

The front of the house seldom has such a range of garden as the back, and as a south or south-west aspect for a garden is preferred, the front usually faces the north or north-east. Here is just the place for some bold masses of evergreens, with a few deciduous trees on the border next the footway. A breadth of lawn, with a bold walk or drive to the house, may be graceful in themselves, but in front of portico, steps, and windows, there is a thin look about it unless shrubs be added. A central compartment of the lawn is usually allotted to hollies, lauristinus, aucuba japonica, tree box, rhododendron, common and Portugal laurel, snow-berry, and such like shrubby-growing evergreens, and these look rich when they get to their full size, and well massed together. The borders, if there are any, may be edged with brickmakers' 'burrs', or, still better, Hogg's tiles, or if a live edging be preferred, common ivy is just the thing. The borders may also be of shrubs, for this front court should present a *fullness* of character, to give the house a substantial aspect, but all must be as trim as the Corporal's boots; no ugly branches straggling out of the line, no rampant growth to shut out light and suggest that you can't afford to keep a gardener. You need not lavish much upon the forecourt, but what you have must be in first-rate order. Then for the border next the public path what can be better than limes, or a pair of the rose-coloured horse-chesnut, with a mixture of almond, lilac, laburnum, lady birch, and acacia, or even apple and pear, all planted out in order, and with a facing of hardy rhododendrons on the side next the house to cheer the eye from the windows with their lovely purple blushes in June and July. The Irish yew and evergreen cypress look well in single plants or pairs on a trim lawn, but all these are better suited for planting apart from clumps and borders, either on grass or gravel, and as they are somewhat gloomy, though very substantial and respectable, must not be chosen without at least some consideration.

In a north or east aspect such a mode of planting a forecourt would afford shelter for flowers, which, of course, should be grown on those borders which were quite within view of the windows. The flowers most suitable for this purpose are the old-fashioned perennials, with a few bedded greenhouse plants, such as geraniums, fuchsias, calceolarias, and myrtles, but high-class flowers or glowing masses of contrasted colours are, in my opinion, quite out of place here, as much so as wire arches and rustic baskets would be. Architectural embellishments are admissible, but they should be sparingly used, for what gives perfect grace and luxuriance to the private grounds has an air of ostentation when exposed to the daily gaze of an 'admiring public'. Spring flowers are unquestionably as much for the joy of the wayfarer as for the proprietor of the house. The invalid who creeps out in spring on the first day that the weathercock indicates a change to a warm quarter, the artizan plodding to work at daybreak or returning fatigued at dusk, nay even the overworked postman and news-boy hurrying to your gate with glad or sad tidings, are entitled to a little of the cheerfulness which you can give them by displaying some of those precious snowdrops, crocuses, hyacinths, and tulips that are never out of place anywhere.

Near the house it is advisable not to plant thickly, though some substantial mansions *do* look all the better for the green boughs that hug them with fond arms; and if strict attention be paid to symmetry of arrangement, some tall laurels or hollies, or, better still, Portugal laurels, to break the angles of the house, and screen the descending pathway where the butcher and baker enjoy a right of way. A gaudy display is most unmeet in a forecourt; so is the gloomy darkness of overgrown and overplanted firs and yews; so is a tangle of forest trees that might have looked very well when young, but which have grown into dimensions that suit them for the park; so is an elaborate pattern of any kind, especially if laid out for flowers with labyrinths of gravel between; so, indeed, is any decided flower garden, even if the place be large; and I would not tolerate, except far away in the country, a display of standard roses or a bed of tulips. Mix the flowers and plant them sparingly, support and vary them with graceful masses of azalea in bloom, iris, agapanthus, delphinium, lily, hydrangea, peony, and chrysantheum, but keep away for use elsewhere every dahlia, holly-hock, rose, verbena, *fancy* calceolaria, pelargonium, and fuchsia that you have, unless you are determined to make a Flower Garden of it, and even *then* dahlias, hollyhocks, and roses, unless on their own roots, will generally be inappropriate. Lastly, balance your work against the outline of the house, for *that* is not to be hidden or treated as if it did not exist.

SHIRLEY HIBBERD, 1856

Villages. – An individual of taste, and of an amiable disposition, who happens to be placed in a village, may, even in the present very imperfect state of things, do much in the way of ornamenting and improving it. We have seen a fine instance of this in the village of Bowness on Windermere. Mrs Starkey, who has ornamented her own house and ground, situated in that village, with many of the finest plants and shrubs, offers seeds of young plants freely to every villager who will plant and take care of them. Mr Starkey has purchased some ground and widened the village street where it was narrow, devoting a marginal space to evergreens and flowers, unprotected by any fence. Mrs Starkey has also planted and carefully trained laurels, box, and holly, against the churchyard wall. In other situations, where laurels would not grow, she has planted ivy; some chimney tops she has ornamented with creepers, and others she has rendered more pictur-esque by architectural additions. Mrs Starkey's own house, which is entered directly from the village street, is ornamented by a veranda which extends its whole length. Independently of woody climbers of the finest sorts, which remain on this veranda all the year, pelar-goniums, georginas, maurandias, lophospermums, and other similar plants, are planted at the base of the trellised supports, and flower there during the summer. At the opposite side of the street is another piece of trellis-work, as the fence to a flower garden: this trellis, when we saw it, was partially covered with purple and white clematis, sweet peas, nasturtium, calampelis, pelargoniums, and georginas. These hung over into the street in profusion; and the gardener assured us that no person, not even a child, ever touched a flower or a leaf. Mr Starkey (a Manchester manufacturer) had not yet arrived there for the season, and the house was in consequence shut up; but of this circumstance the villagers took no advantage. In the gardens of this village, and in part also in those of Ambleside and Grasmere, may be seen many of the new potentillas, geums, lupines, clarkia, &c.; and against the walls, kerria, Cydonia japonica, China roses of different sorts, clema-tis, and other climbers are not uncommon. The village of Bowness affords a proof that, when the public are treated with confidence, they will act well in return; and that, notwithstanding what has been said of the rudeness of John Bull, he will, when treated like the French and Germans, become as considerate and polite as they are. It is true, the working inhabitants of London and of manufacturing towns cannot be expected all at once to pay the same respect to flowers as the inhabitants of Bowness; but time will remedy this evil.

JOHN CLAUDIUS LOUDON, 1831

Town gardens. – Under this denomination we include the gardens and grounds attached to houses in streets, and also the gardens belonging to persons living in towns, but which are detached from their houses; the latter being gardens of culture only.

The detached town gardens are situated in the suburbs of towns, generally collected together, and separated by hedges. There are upwards of two thousand of such gardens in the neighbourhood of Birmingham, a considerable number at Wolverhampton, some at Dudley and at Manchester, and a few even in the neighbourhood of that stationary town Buckingham.

It is not uncommon for single men, amateurs, clerks, journeymen, &c. to possess such gardens, and to pass a part of their evenings in their culture. In one of these gardens, occupied by Mr Clarke, chemist and druggist, Birmingham (the inventor of *Clarke's Marking Ink*), we found a selection of hardy shrubs and plants which quite astonished us.

JOHN CLAUDIUS LOUDON, 1831

Cottage gardens. – By these we understand the gardens attached to cottages in villages, or to the humbler class of dwellings scattered through the country. In the agricultural district from London towards Warwick they are small and poorly cultivated, in comparison with those around Birmingham and the other manufacturing towns. The cause is too obvious to require explanation. We are not sure, however, that the culture of flowering shrubs against the walls of cottages is so general in the manufacturing as in the agricultural districts. We expected greater progress to have been made among the gardens of the miners in Derbyshire, where we found, indeed, in Middleton Dale, and at Castleton, some cottages without gardens. We recommend this subject to the Duke of Devonshire's agents, and to Mr Paxton, who might distribute plants and seeds among them.

JOHN CLAUDIUS LOUDON, 1831

Before we go inside our house, nay, before we look at its outside, we may consider its garden, chiefly with reference to town gardening; which, indeed, I, in common, I suppose, with most others who have tried it, have found uphill work enough – all the more as in our part of the world few indeed have any mercy upon the one thing necessary for decent life in a town, its trees; till we have come to this, that one trembles at the very sound of an axe as one sits at one's work at home. However, uphill work or not, the town garden must not be neglected if we are to be in earnest in making the best of it.

Now I am bound to say town gardeners generally do rather the reverse of that: our suburban gardeners in London, for instance, oftenest wind about their little bit of gravel walk and grass plot in ridiculous imitation of an ugly big garden of the landscape-gardening style, and then with a strange perversity fill up the spaces with the most formal plants they can get; whereas the merest common sense should have taught them to lay out their morsel of ground in the simplest way, to fence it as orderly as might be, one part from the other (if it be big enough for that) and the whole from the road, and then to fill up the flower-growing space with things that are free and interesting in their growth, leaving Nature to do the desired complexity, which she will certainly not fail to do if we do not desert her for the florist, who, I must say, has made it harder work than it should be to get the best of flowers.

WILLIAM MORRIS, 1882

Whatever builders may say about usage, and expense, and doing as their fathers did before them, it must be admitted that a fundamental principle of taste is violated when we give our houses handsome frontages to the public, and reserve for our own daily contemplation from the garden, nothing but bare walls and plain windows, and oblique chimneys rising from a basement of ugliness . . . It is as bad as for a man to appear in society with a showy vest and faultless collar, but with soiled fustian at his back, because, forsooth, you are not expected to address him from behind . . .

SHIRLEY HIBBERD, 1856

For some years past I have been counting myself lucky to have several stretches of north wall of various heights, which, after successfully nursing a cleaning crop of potatoes, are now proving not much less useful in their own sphere than south walls, and rather more useful than similar walls facing east, or perhaps I should be honest and write 'slightly west of north' and 'slightly north of east'. I could never quite understand why this garden was slightly askew by the pole-star, until, happening one morning this summer to be about rather early, I found another wall, which to my knowledge runs dead east and west, bathed in early sunlight. The slight skewing of these old walls cuts down the dangers of the early sun, and prolongs the exposure to the evening warmth.

HUMPHREY DENHAM, 1940

As to the other desirable qualities [of a garden] – animation, variety, mystery – I would base my garden upon the model of the old masters, without adopting any special style. The place should be a home of fancy, full of intention, full of pains (without showing any); half common-sense, half romance; ... it should have an ethereal touch, yet be not inappropriate for the joyous racket and country cordiality of an English home. It should be ... something that would challenge the admiration and suit the moods of various minds; be brimful of colour-gladness, yet be not all pyramids of sweets, but offer some solids for the solid man; combining old processes and new, old idealisms and new realisms; the monumental style of the old here, the happy-go-lucky shamblings of the modern there; the page of Bacon or Temple here, the page of Repton or Marnock there. At every turn the imagination should get a fresh stimulus to surprise; we should be led on from one fair sight, one attractive picture, to another; not suddenly, nor without some preparation of heightened expectancy, but as in a fantasy, and with something of the quick alternations of a dream.

Your garden, gentle reader, is perchance not yet made. It were indeed happiness if, when good things betide you, and the time is ripe for your enterprise, Art 'Shall say to thee / I find you worthy, do this thing for me.'

JOHN D. SEDDING, 1890

As I write this, on June 29, it's about time for another summer storm to smash the garden to pieces, though it may hold off until the phlox, tomatoes, daylilies, and zinnias are in full sway.

I detect an unwholesome strain in gardeners here, who keep forgetting how very favorable our climate is, and who seem almost on the verge of ingratitude. Disaster, they must learn, is the normal state of any garden, but every time there is wholesale ruin we start sounding off – gardeners here – as if it were terribly unjust. Go to any of those paradise-type gardens elsewhere, however, and see what they put up with in the way of weather, and you will stop whimpering. What is needed around here is more grit in gardeners.

Now I guess there is no garden in the world more dreamworthy than the one at Tresco Abbey in the Scilly Isles. It rarely approaches freezing there, off the mild coast of England, and wonders abound. Palms grow luxuriantly against soft old stonework, medieval in origin, and there is hardly an exotic rarity of New Zealand or South Africa or Madeira that does not flourish. And yet they can have their daffodils, too, for it never gets hot in those islands, either; and if you view such a garden in the long slanting light of a summer's late afternoon, you will think you have got to heaven in spite of yourself. Indeed, almost any garden, if you see it at just the right moment, can be confused with paradise. But even the greatest gardens, if you live with them day after decade, will throw you into despair. At Tresco, that sheltered wonderland, they wake up some mornings to discover 500 trees are down – the very shelter belts much damaged. The cost of cleanup is too grim to dwell on, but even worse is the loss not of mere lousy Norway maples, but of rare cherished specimens that were a wonder to see in flower.

Or there may be – take the great gardens of Gloucestershire – a drought, and the law forbids you to run the hose. Not just a little dry spell, either, but one going on month after month. There you sit in your garden, watching even the native oaks dry up, and as for the rarities imported at such cost, and with such dreams, from the moist Himalayas, the less said of their silent screams the better.

Or take another sort of garden, in which the land to begin with is a collection of rusting bedsprings and immortal boots. Old shoes simply do not rot, in my opinion, but just stay there forever. The chief growth the gardener finds (I am speaking now of the great garden of Sissinghurst in Kent) is brambles and bracken and dock, maybe broken up by patches of stinging nettles. Amenities include the remains of an old pig sty. You convert it, let's say, into one of the sweetest gardens of the world, with roundels of clipped yew and a little alley of lindens, rising over a wide walk, almost a terrace, of concrete cast in big blocks (not one in a thousand knew it was concrete) with spaces for a riot of primroses and spring bulbs, bursting out everywhere in lemon and scarlet and gentian and ivory. The lindens all die. The pavement has to be replaced. The primroses start dying out – they develop a sickness, they wear out the soil, and no mulches of manure, no coddling of any sort will preserve them. So you grub out the dying and start anew with something else.

Wherever humans garden magnificently, there are magnificent heartbreaks. It may be forty heifers break through the hedge after a spring shower and (undiscovered for many hours) trample the labor of many years into uniform mire. It may be the gardener has nursed along his camellias for twenty-five years, and in one night of February they are dead. How can that be? Well, it can be. You have one of the greatest gardens of the Riviera, and one night the dam of the reservoir breaks. The floor of the house is covered with a foot of mud once the water subsides. The reservoir was built at endless labor and cost, since the garden would die without water from it. And now it is gone, and in the flood everything has gone with it. Be sure that is not the day to visit that great garden.

I never see a great garden (even in my mind's eye, which is the best place to see great gardens around here) but I think of the calamities that have visited it, unsuspected by the delighted visitor who supposes it must be nice to garden here.

It is not nice to garden anywhere. Everywhere there are violent winds, startling once-per-five-centuries floods, unprecedented droughts, record-setting freezes, abusive and blasting heats never known before. There is no place, no garden, where these terrible things do not drive gardeners mad.

I smile when I hear the ignorant speak of lawns that take 300 years to get the velvet look (for so the ignorant think). It is far otherwise. A garden is very old (though not yet mature) at forty years, and already, by that time, many things have had to be replaced, many treasures have died, many great schemes abandoned, many temporary triumphs have come to nothing and worse than nothing. If I see a garden that is very beautiful, I know it is a new garden. It may have an occasional surviving wonder – a triumphant old cedar – from the past, but I know the intensive care is of the present.

So there is no point dreading the next summer storm that, as I predict, will flatten everything. Nor is there any point dreading the winter, so soon to come, in which the temperature will drop to ten below zero and the ground freezes forty inches deep and we all say there never was such a winter since the beginning of the world. There have been such winters; there will be more.

Now the gardener is the one who has seen everything ruined so many times that (even as his pain increases with each loss) he comprehends – truly knows – that where there was a garden once, it can be again, or where there never was, there yet can be a garden so that all who see it say, 'Well, you have favorable conditions here. Everything grows for you.' Everything grows for everybody. Everything dies for everybody, too.

There are no green thumbs or black thumbs. There are only gardeners and non-gardeners. Gardeners are the ones who ruin after ruin get on with the high defiance of nature herself, creating, in the very face of her chaos and tornado, the bower of roses and the pride of irises. It sounds very well to garden a 'natural way'. You may see the natural way in any desert, any swamp, any leech-filled laurel hell. Defiance, on the other hand, is what makes gardeners.

HENRY MITCHELL, 1981

Because Art stands, so to speak, sponsor for the grace of a garden, because all gardening is Art or nothing, we need not fear to overdo Art in a garden, nor need we fear to make avowal of the secret of its charm. I have no more scruple in using the scissors upon tree or shrub, where trimness is desirable, than I have in mowing the turf of the lawn that once represented a virgin world.

JOHN D. SEDDING, 1890

The qualities of water have to be classified and studied or the keen edge of enjoyment is blunted by muddle, for it can be appreciated in so many different ways, be it hosed through dusty, city streets on an August day, or drawn up from holy wells to sprinkle over pilgrims, or used as the centre piece of a desert *glorietta*. It is essential for a truly epicurean enjoyment of water that its best characteristics be categorized and that those with sufficient force be kept separate from each other or, like Petruchio and Kate, only married with the greatest care. This principle is not as obvious as it seems. Hotchpotch water gardens where the designer has aimed for both simplicity and stimulation and managed to achieve neither are fairly common. And there are some professional designers who evidently feel they must justify their fat fees by recommending quite unnecessary and often elaborate alterations, going so far sometimes as to straighten or even pipe a natural stream to suit a formal and unsuitable scheme. Or, alternatively, and equally perversely, they will corkscrew an open drainage dyke at great expense in the hope that nitwit visitors to the garden will think it a natural stream. If we make our own plans, and gardening, being such a highly individual art, is often better done by amateurs, we shall probably base them on any assets that happen to be handy, and so bind ourselves naturally to the canon that a water garden ought to suit its setting. This, at any rate, is the convenient rule unless we have a considerable genius for improvisation and organization, vast space, immense money-bags, and a Pharaoh's task force to do it.

TYLER WHITTLE, 1969

*C*arpet Bedding. By this term is meant regular and formal arrangements of dwarf *foliage* plants, both tender and hardy, either alone or in combination with flowering plants. The popularity of this style of bedding has excited a corresponding amount of adverse criticism from parties who can see nothing to admire in anything that does not coincide with their own notions of beauty. It would be uncharitable to accuse such people of want of taste, but their taste is certainly of a curious order, where no 'keeping' of lawns, and the just allowing of all flowers to grow at random, or as they themselves prefer to put it, 'at their own sweet will', is the professed model of their gardening. This we call at least as great an extreme in the opposite direction, and of the two prefer carpet bedding, which it is possible to carry out with far less of formality than is generally done.

The term of *carpet bedding* is not a good one, but doubtless originated from the table-like flatness which the arrangements were made to assume when the style was first introduced, a mode still practised by some. But a more graceful style is gradually springing up, the stiffness or formality being broken at regular intervals with standard graceful-foliaged plants, so that the term 'carpet' is no longer appropriate, and a far better term coming into use is *panel gardening*.

WILLIAM WILDSMITH, *c.* 1890

There is always hope for the man who loves flowers, and I for one will never believe that the bottom has fallen right out of England so long as there are gardens, and men to tend them faithfully. First the planning, then the making and planting, finally the blossoming. There are many ways of making a garden, or rather there are many styles of garden, formal and informal; and the elements of a garden, the plants themselves, are legion. Now these are the bare elements of a garden: form and colour, scent and sound, and the surprise of contrast; and it needs an elegant blending of all to make a perfect garden.

Anyone who seeks that 'peace which passeth all understanding' or, as the Buddhist calls it, Nirvana, might well turn to his garden to distract his mind from the daily anxieties of life. We all love flowers for their graciousness of form, no less than for their delicious colours, and sweet fragrance; and there is nothing which gives so great a return for the labour and money expended on it as a small garden. Thanks to quick transport from the scene of our daily labours, few of us are without some sort of garden, if we can afford to live only a few miles out in the country; even for those who are forced to live in the city, there are public parks where they can enjoy vicariously the gardener's craft. When the city man comes home, tired after his day's work, it is to his garden that he turns, not only for recreation but for comfort. Sitting outside after supper one warm June night he watches the shadows lengthening on the lawn as the sun drops slowly in the west; a great peace descends upon him. We can picture the scene, as he sits there drinking in the richness of the earth, while the outlines of the trees melt into the lilac dusk. One by one the colours are eclipsed, first the reds, then the blues, last of all the yellows. But the scent of the rose lingers. A faint breeze stirs the leaves; and a bat wheels swiftly overhead, and is gone like a spirit. Now it is dark; and presently without warning the trees begin to cast shadows anew, faintly at first and on the far side of the lawn. A rich glow suffuses the east; the full moon has risen. Thus he sits, bathed in the golden radiance; and as he dozes his troubles slip from him like a garment, and he dreams of the flowers which have sprung up at the touch of his hands. There they stand in royal array, a tribute to his skill and patience; Roses, Irises, Larkspurs, Paeonies, Lupins, and many more, in tumultuous colours and shapes. He has watched over them in their infancy, tended them in sickness, gloried over them in their prime. He has made of them that worth while thing, a garden; and in blossoming time the garden has been as balm to his soul. So, even in these hard times, let us not neglect our gardens, but rather sacrifice much else, that we may enjoy the purest and most exquisite pleasure in life: the yield of mother earth in all its strange forms. Truly the cultivation of flowers is something more than a luxury; it is a religion.

FRANK KINGDON WARD, 1935

While the popularity today of architectural gardens with a succession of hedged or walled rooms is partly due to the urban background of their owners, who feel more secure in an extension of the house into the garden than in an extension of nature into the garden, they have another appeal too. That is, the provision of shelter in the garden's various compartments for different kinds of plants. This is surely an admirable practice, unless it degenerates into housing a *collection*. Innumerable different species, varieties, and cultivars, without botanical or aesthetic significance, are arranged like so many stamps in an album to make up that contemporary horror, the 'plantsman's garden'.

DOUGLAS SWINSCOW, 1992

Any gardening is better than none, but some ways of gardening are better than others. Better gardeners seem to sense which plant is best, how to find the best oak or primrose and how to place it for the best impression. Their roses are musk roses, the silver pink Felicia, which needs no pruning. Their catmint is the form called Six Hills Giant which bushes out in a border to a height of three feet and flowers in a wonderful shade of violet-blue. Their bedding salvias are the electric blue Salvia patens, not the harsh and hateful Scarlet Bedder. Their only asters are six inches high, the half-hardy pappei which they grow for their garden urns. They sow no marigolds, except for the peculiar Tagetes minuta, a four-foot-high wild flower whose seed can be bought through specialists' societies. Sown under glass, it is planted out in June wherever better gardens are bothered by couch-grass and ground-elder. These weeds cannot bear the Tagetes' smell or the presence of its roots. This tall wild marigold gradually drives them out.

I know these better gardeners, you may be thinking. They begin with all the advantages, old brick walls and a staff of two or three. Their homes have been gardened for centuries, heavy with holm oaks and bulging yew hedges which set the guide-books cooing, like the resident flocks of white doves. 'Nestling deep in the complicated headwaters of the River Itchen, which rises at nearby Cheriton, Tichbourne possesses a garden of considerable charm from which the famed Tichbourne Dole is annually distributed . . .' But if this is better gardening, I belong at the further end of the queue.

The less promising your site, the more you are in my thoughts. My garden slopes north in three separate sections of a third of an acre, one a rough bank, the other a walled and terraced enclosure, the third, in theory, a vegetable garden and orchard. The first is beset by a neighbouring dentist, young and balding, with more pruning shears than taste, the second by bindweed which has lodged in all the boundaries' stone walls, the third by the deep and intractable roots of Mare's Tail, a primeval weed which resists almost every poison and roots beyond all human excavation. There is no good to be said of it, except that a Yorkshire reader once sent me a recipe for boiling its young shoots and straining them into a pungent thin soup. The soil is limey and heavy. I shaped the ground into its plan of banks and terraces with the help of a mechanical digger some five years ago. When my back was turned, the driver buried most of the top soil and left me with a surface of clay, stones and cracked Victorian china which had never expected to see the light again. On this corner of the heap, I have learnt my lessons.

The most general lesson, however, applies to any garden, large or small. Wherever you choose a plant, the choice can be made for better or worse. In one and the same family lie good plants and very good plants. By watching, reading and experimenting, you begin to learn those which are better than others. The smaller the garden, the more precious the space and the more urgent it is to choose the best. Hence, so many small gardens have an originality which the great set-pieces of the Itchen valley sometimes lack.

ROBIN LANE FOX, 1982

A smooth, closely shaven surface of grass is by far the most essential element of beauty on the grounds of a suburban home. Dwellings, all the rooms of which may be filled with elegant furniture, but with rough uncarpeted floors, are no more incongruous, or in ruder taste, than the shrub and tree and flower-sprinkled yards of most home-grounds, where shrubs and flowers mingle in confusion with tall grass, or ill-defined borders of cultivated ground. Neatness and order are as essential to the pleasing effect of ground furniture as of house furniture. No matter how elegant or appropriate the latter may be, it will never look well in the home of a slattern. And however choice the variety of shrubs and flowers, if they occupy the ground so that there is no pleasant expanse of close-cut grass to relieve them, they cannot make a pretty place. The long grass allowed to grow in town and suburban grounds, after the spring gardening fever is over, neutralizes to a certain degree all attempts of the lady or gentleman of the house to beautify them, though they spend ever so much in obtaining the best shrubs, trees, or flowers the neighbors or the nurseries can furnish. It is not necessary to have an acre of pleasure ground to secure a charming lawn. Its extent may always be proportioned to the size of the place; and if the selection of flowers and shrubs and their arrangement is properly made, it is surprising how small a lawn will realize some of the most pleasing effects of larger ones. A strip twenty feet wide and a hundred feet long may be rendered, proportionally, as artistic as the landscape vistas of a park.

FRANK SCOTT, 1870

Enter a Gardener *and two* Servants.

GARDENER   Go, bind thou up yon dangling apricocks,
Which, like unruly children, make their sire
Stoop with oppression of their prodigal weight:
Give some supportance to the bending twigs.
Go thou, and like an executioner,
Cut off the heads of too fast growing sprays,
That look too lofty in our commonwealth:
All must be even in our government.
You thus employ'd, I will go root away
The noisome weeds, that without profit suck
The soil's fertility from wholesome flowers.
FIRST SERVANT   Why should we in the compass of a pale
Keep law and form and due proportion,

Showing, as in a model, our firm estate,
When our sea-walled garden, the whole land,
Is full of weeds, her fairest flowers chok'd up,
Her fruit-trees all unprun'd, her hedges ruin'd,
Her knots disorder'd and her wholesome herbs
Swarming with caterpillars?
   GARDENER   Hold thy peace:
He that hath suffer'd this disorder'd spring
Hath now himself met with the fall of leaf;
The weeds that his broad-spreading leaves did shelter,
That seem'd in eating him to hold him up,
Are pluck'd up root and all by Bolingbroke;
I mean the Earl of Wiltshire, Bushy, Green.
   FIRST SERVANT   What! are they dead?
   GARDENER   They are; and Bolingbroke
Hath seiz'd the wasteful king. O! what pity is it
That he hath not so trimm'd and dress'd his land
As we this garden.

WILLIAM SHAKESPEARE, *c*.1595

ON A FINE CROP OF PEAS BEING SPOILED
BY A STORM

When Morrice views his prostrate peas,
   By raging whirlwinds spread,
He wrings his hands, and in amaze
   He sadly shakes his head.

'Is this the fruit of my fond toil,
    My joy, my pride, my cheer!
Shall one tempestuous hour thus spoil
    The labours of a year!

Oh! what avails, that day to day
    I nursed the thriving crop,
And settled with my foot the clay,
    And reared the social prop!

Ambition's pride had spurred me on
    All gard'ners to excell.
I often called them one by one,
    And boastingly would tell,

How I prepared the furrowed ground
    And how the grain did sow,
Then challenged all the country round
    For such an early blow.

How did their bloom my wishes raise!
    What hopes did they afford,
To earn my honoured master's praise,
    And crown his cheerful board!'

Poor Morrice, wrapt in sad surprise,
    Demands in sober mood,
'Should storms molest a man so wise,
    A man so just and good?'

Ah! Morrice, cease thy fruitless moan,
    Nor at misfortunes spurn,
Misfortune's not thy lot alone;
    Each neighbour hath his turn.

Thy prostrate peas, which low recline
    Beneath the frowns of fate,
May teach much wiser heads than thine
    Their own uncertain state.

The sprightly youth in beauty's prime,
  The lovely nymph so gay,
Oft victims fall to early time,
  And in their bloom decay.

In vain th' indulgent father's care,
  In vain wise precepts form;
They droop, like peas, in tainted air,
  Or perish in a storm.

HENRY JONES, 1749

I generally arrange to be absent from my garden in mid and late June, for I am one of those badly finished off persons whose mucous membrane never got the last coat of paint, or the right tempering or hardening, or whatever was needful to enable it to resist the irritation of grass pollen that is called Hay-fever. I believe I have tried every remedy that has been put on the market, and though some alleviated my particular forms of sneezing and eye-swellings, none made me feel well enough to be happy. I objected all along to have my nose cauterized, believing it dulls one's power of scent, which means so much to me that I would far rather snuffle and sneeze for one month and be able to smell clearly and keenly the other eleven than be robbed of any olfactory powers; and before the days of anti-toxins and injections I discovered so pleasant a cure that Hay-fever has become quite a valuable asset in my life scheme, for I must, '*absobally-lutely must*', as Grossmith used to say, carry my poor nose away from the flowering grass meadows to Alpine heights where a breeze blows off the snow. Once I reach an altitude of 3,000 feet I am cured, and the sight of *Poa alpina* in its viviparous state by the side of a road assures me it is safe to draw in the breeze with expanded nostrils. A sea voyage might take me into a pollenless region; but I hate the sea as much as I do Hay-fever, and, on the other hand, am glad of so good an excuse to get away to Nature's own rock garden at the very time it is making its bravest show, and looking as if it were doing so on purpose to please me and invite me to help myself to whatever I like.

E. A. BOWLES, 1914

If art historians are in general agreement that a urinal signed by Marcel Duchamp constitutes a landmark in twentieth-century sculpture, who am I to disparage the gentleman in our neighbourhood whose front garden features a row of toilet bowls? (Planted up, I hasten to add, with pelargoniums.) In both instances I must once again confess to being somewhat baffled by the world around me, and even further inclined to the theory that we all live in an extended lunatic asylum: a theory which has the merit of rationalizing a good deal of otherwise inexplicable human behaviour.

Since there seems to be no clear definition nowadays of what is sculpture – or more accurately, no exclusion clauses proscribing what is *not* sculpture – to write about sculpture in the garden is a task fraught with conceptual difficulties; preferable to discuss garden ornament, which may be taken to include anything from a pagoda to a *pithos* (but ignoring gnomes, which have already been given more press than they deserve).

We habitually think of ornament as something which is introduced into the existing horticultural scheme of things, but historically the reverse appears to have been the case. In the beginning was the totem pole, the stone altar, a tastefully arranged pile of enemy skulls, or some branches bedecked with votive offerings. The earliest sacred groves were naturally-occurring groups of trees (or whatever plants conveniently served the purpose), and only later were such sites deliberately landscaped. In other words, the ornaments were first put in place and gardening subsequently evolved round them. You may consider the implicit assumption that the sacred grove is the direct ancestor of Vita and Harold's Sissinghurst to be unwarranted, but I will defend it.

Someone once remarked that the English have transferred their sense of religious observance to the cultivation of their gardens – and why not? Many of us may now be godless, but we still have to cater for the needs of the spirit. Another phenomenon which I find significant in this respect is the vogue for barbecues. In the British climate, where one can only dine al fresco in exceptional summers, the barbecue must be viewed as a piece of ritual, rather than practical, equipment. The close similarity to a primitive altar confirms that barbecuing is an atavistic activity reminiscent of sacrificial rites (and the culinary results are often not much better).

Ornament, therefore, is aboriginal, and we have no need to justify its inclusion in the garden. (Quite the opposite, in fact.) . . . The choice of objects which could conceivably be incorporated into a garden design is so wide that it would be prudent to dismiss some of the sillier examples out of hand in a short essay such as this. The Wishing Well, for instance: anyone who believes that this structure imparts an element of idyllic rural charm has never lifted a bucket of water. For the villager (that is to say, the village woman or child) without the luxury of water on tap, the well represents a source of unending drudgery and the occasional bout of typhoid fever.

On the other hand, there must be countless objects whose potential as garden ornament has yet to be explored. Not far from here (just outside Athens) an ancient village bus stands on the edge of an abandoned vineyard. Windowless (recalling the sightless eyes of old Greek statues), its blue and white livery flecked with ochrous rust, it no doubt once served as a shed during the grape harvests. Doubly retired, now in spring and early summer it is half-submerged amid mounds of yellow *Chrysanthemum coronarium* mixed with the white of *C. coronarium discolor*, which foam around it as if it had been halted by a golden flash flood. While I am fond of this feature of our local landscape, it is not so easy to determine what distinguishes this derelict vehicle from the numerous other eyesores of rubbish dumped hereabouts. I hesitate to advocate it as a paragon of ornament, since your garden could well end up resembling a cross between a transport museum and a pop festival car park.

One cannot discuss garden ornament without reference to classical statuary and other antique whatnots. Such *objets d'art* are not only *toujours à la mode* but have acquired the *cachet* of being *de rigueur*. Faced with so many linguistic borrowings it is only natural to entertain some doubts on this score. If anyone could explain to me the relationship between a recent copy of a nineteenth-century lead reproduction of a third-century BC marble Attic sarcophagus, and the freshly landscaped property behind a row of restored paupers' cottages in Kent, I would be most grateful. Indeed, it is not so much a question of what you put, as where you put it. The most magnificent of stone troughs would do nothing to enhance a garden on a modern housing estate. (As a matter of fact, I cannot think offhand of anything that *would*.) Real stone and real stonemasonry would simply put the plastic 'Georgian' windows to shame.

And what of a naked Aphrodite in marble – or an equally unclothed Pan (with flute) in bronze? By what logic do English garden-makers expose these Mediterranean deities to the risk of hypothermia, so close to the Arctic Circle? Curiously, we never see representations of members of our own northern pantheons – though Freyja or Loki would have been more sensibly dressed in wool and linen. Our empire-building forefathers, of course, carried Olympian god and sylvan nymph into boreal exile because – like the Romans who plundered Greek culture before them – they scored high in civil engineering and administration, but were dunces when it came to the fine arts. Not wishing to be labelled barbarians, they were diligent in their efforts to fill the gaps in their education (and gardens). They took their Classics seriously: some were members of The Athenaeum and one or two could even parse a Latin sentence. Anyone who could afford it dedicated a ruined temple to Zeus at the head of an artificial lake. But it is the continuing nostalgia for the distant past which is perplexing. Since the Classics (along with anything else which might have had a civilizing influence on the young) have been dropped from the school curriculum, and the majority of today's purchasers can thus not claim even a passing acquaintance with the ancient world, one wonders what impels them to acquire a pastiche of Praxiteles?

The ultimate irony of this situation is that while *we* stand before these alien images with that sense of obligatory reverence which comes with ignorance, Greeks of the late Classical period had themselves ceased to take their myths at all seriously.

But there you have it: every garden journal now carries its quota of advertisements for antique garden whatsits, which have to be mass produced in order to supply the appetite of the market. Mass production has a long history, and the ancient Greeks were not above churning out standardized ceramic effigies in quantity. This, however, does not make repetition any more appropriate in the realm of art. In the words of another author whose name escapes me: 'The first man who said "My love is like a rose" was a poet. The second was an idiot.' Likewise, a pottery hedgehog balancing an apple on its snout could be quirky and diverting when first encountered in a shrubbery; subsequent sightings will lack the ingredient of surprise, giving you time to reflect on the banal inanity of the thing. One might think that this prospect would be enough to deter anyone from installing ornaments known to be in other people's gardens, but the opposite seems to be true. If *you've* got one (and even more urgently, if everybody else has got one), then *I've* got to have one, too. How sad.

Meanwhile Aphrodite, shivering in the icy drizzle, is bemoaning her fate to the rhododendrons around her. From the ensuing conversation it transpires that they have also been transported far from their native land. Other plants have similar stories to tell – for the garden is a veritable Babylon. Very soon the ornaments swell the lament of forced expatriation, the Italian lemon pot, chinoiserie bench, Indian lattice summer house and Japanese stone lantern between them creating such a racket that the Koi carp in the pool can hardly make themselves heard.

Thankfully, not every garden is as eclectic as I have suggested here. We are more likely to visit Italian gardens, Japanese gardens, and so on. There are even English gardens – but a tradition of indigenous ornament seems to have stopped short at the bird-table and nesting-box (unless we include the clothes-line). Otherwise, everything is imported. To be perfectly fair, this applies not only to Britain: somewhere in France a garden party is taking place in a Persian kiosk, while in Spain the *sangria* is being enjoyed on all-American cedar decking. Wherever you go, anything goes.

DEREK TOMS, 1995

Though artefacts may keep us alert with their ingenuity, a deeper undercurrent of expectation binds us more firmly to our own garden. We live for the future; dig, sow, weed, hoe, stake, and prune all for the future. We may stand and admire the garden for a few moments, but soon we start imagining what it ought to look like in a month's or a year's time. Expectation is what gives us the face of contentment as we work in it through sun and rain. We labour in accordance with the needs of the garden, and our life joins there with the life of nature. The garden has a future, and so have we.

DOUGLAS SWINSCOW, 1992

CHAPTER TWO

People in the Garden

At the age of five I was allotted some twelve square feet of soil for a garden. This had been dug and raked to a fine tilth before being handed over, and no Capability Brown ever experienced a greater thrill than I in planning a new estate. The family had recently visited an old property, where a small enclosure took my fancy to the exclusion of all else, causing me to linger in what I ignorantly thought had been described as a 'not-garden', although it looked very much an 'is-garden' to me.

After this, nothing would do for my plot but to copy the novel pattern on home ground as best I could. Laboriously I scratched out a circle surrounded by squares and oblongs, making narrow trenches in place of the original hedges. My elder brother sniffed critically at the lopsided circle and crooked rectangles, saying the plan reminded him of our bathroom linoleum. He was being bossy on account of his thirteenth birthday and a first pair of long trousers. Brother Ben, his junior by three years, was far kinder. Bending over my plot, he evened up the circle by means of a trowel tethered to a central stake, a practical home-made compass.

Next, there were seeds to be bought. In a shop stacked with bins of chicken-feed and meal, our friend Miss Kirkby kept neat trays of seed packets with pictures on them of how their contents were supposed to turn out. I believed implicitly in the truth of this gorgeousness. Before leaving home I had unlocked my savings box and withdrawn the entire capital, thirteen copper pennies with the face of King George V on them. Knowing that seeds cost twopence a packet, I was able to work out that six would leave me with one penny over. A sneaking idea that the nice shopkeeper might let me have seven for thirteen pence was kept to myself.

Accompanied by Ben I walked, ran and skipped downhill to the shop, among whose bins and sacks even the small squares of dog biscuit looked tempting. Ben called me to order, and together we studied the seed tray. First I picked out eschscholtzia, for I was proud of knowing how to spell this alarming name, taught by my gardening grandmother, and I always had enjoyed pulling little dunce's caps off her plants. She also grew clarkia and godetia, so I picked them out. Then I must have love-in-a-mist, and the strange brownish plumes of mignonette for their lovely scent. A packet of cornflowers swallowed up my shilling, and only one penny remained for Shirley poppies and nasturtiums. I was close to tears when Ben pulled a silver threepenny-bit from the pocket of his shorts. 'Here you are,' he said. 'Oh, can you *really* spare it?' 'Of course I can.' So the deal was done.

Joy of possession soon turned to dismay when I wondered how to divide the supply evenly among the marked spaces without overcrowding the plants or leaving gaps between them. Somehow Ben grasped the problem, and solved it with ease. Scrounging a roll of kitchen paper, he spread a sheet on our playroom table and drew with ruler and compasses a scale plan of my plot. Then he worked out how best to distribute the contents of eight seed-packets, which he carefully numbered. His calculations were beyond me, but my nimble fingers were useful for setting groups of seed into hollows in a paintbox palette. Next, Ben fashioned a collection of marker sticks, numbered to match the packets.

After our midday dinner we went out to copy Ben's plan on the ground. It took us all the afternoon, by Ben's watch, but he said the idea had worked out well, and his marker sticks were 'a good wheeze'. He knew exactly where the different groups of my chosen seeds had been sown. Now, of course, there would be many weeks of waiting before they came into flower. I was quite ready for this, having helped Granny to sow annuals when I was only four, last year. Happily I went off to bed and fell asleep to the hooting of owls under a bright moon. No hint of tomorrow's rude shock disturbed my dreams.

It was Ben who broke the shattering news, after we had finished breakfast. Some animal, he said, *possibly* Bella our family cat, had used my plot overnight for its own purposes, scratching the grooved pattern all abroad and leaving our carefully sown seeds in a terrible mess which nobody could ever sort out. I was too shocked to go and look at the wreck of my poor garden. Granny, on a visit to us, having breakfasted in her own room, took charge of the situation. 'My dear child,' she said to me, 'don't give it another thought. Those annuals will look even better mixed up together, you'll see.' Something about her quiet face and the lace cap perched upon white hair reminded me of a marble monument in the church. She just had to be listened to and believed. 'Remember that patchwork quilt which you liked so much when you stayed with me? Well, you are going to have a patchwork garden, and I'm sure it will be quite lovely; much prettier and more natural than a stiff old knot-garden.'

As the season advanced and seedlings appeared, I did manage to thin out certain groups and make a few transplants into gaps left by Bella, but no trace of the original design remained. Everything grew well, and by the end of a warm July, with gentle refreshing showers, there was a grand display of blossom. On a Sunday morning I walked round the whole garden hand-in-hand with my father. Stopping at my little plot he exclaimed, 'This is the best show of the lot! It reminds me of a wild flowery meadow beside our rectory when I was a boy.' So I hadn't even made a wild garden, just a piece of meadow-land. Because Father evidently preferred it that way, the plot pleased me, too. Looking back down the long tunnel of time, I can see now that my original misspelling wasn't far out, after all.

DAWN MACLEOD, 1994

Magpies were manly, for Jones the Prefect kept one, and guinea-pigs were quite genteel; but we might not even speak of flowers. They were considered to be beneath the dignity of gentlemen who would be nine years of age next birthday. It was quite legitimate to make toffy, and it was honourable to play at horses; but when Simpkins *minimus* bought a fuchsia there was one wild howl of scorn. Entomology had ardent friends: butterflies were extensively impaled on cork; silkworms were openly maintained; cockchafers were diligently harnessed to elegant landaus of walnut-shell, with buttons of bone for wheels; frequent trials of strength and science were very generally encouraged between the various species of the genus beetle, and a terrific duel between a 'Soldier' and a 'Sailor', which took place in a large pill-box, was the thrilling topic of 'the half'. There were collections of eggs, shells, seals, and autographs (I myself presented one of the juniors with a unique assortment of the latter, comprising scores of distinguished heroes, from Alfred the Great to Bendigo, and all written, like Dr Johnson's Dictionary, with a single pen), but not a boy of us dared to bring home a posy. Any such pusillanimous proceeding would infallibly have evoked from our disgusted community those epithets, so awful to the boyish soul, viz., 'softy', 'nincompoop', and 'mollycoddy'. We trembled to remember those dark ages when we had loved the wild rose and the honeysuckle, when we had filled our small fists with violets, made golden balls of the cowslip, decked ourselves with daisy-chains, and when we were never weary of the tiny garden which was called our own.

S. REYNOLDS HOLE, 1909

Somewhere there is a photograph of me, aged about four years, in the act of moving a toy wheelbarrow. It is obvious that gardening attracted me from early days but the matter was not really finalized until my Godfather gave me a large-flowered fuchsia at the age of six. It was this – and also the Godfather who was a great gardener – which set me on an earthy career. The fuchsia was planted in the small patch of ground which I called 'my garden', and, being covered every winter with a pile of ashes, survived for several years. The fact that my little patch was under an apple tree and beyond the laurel hedge (which marked the end of the garden proper and screened an area for rubbish and bonfires) did not dim its lustre in my eyes. Imagination lent it every quality and dimension.

When we are young every obstacle is surmounted by the imagination; and if the imagination is fertile, fertilized ideas have no bounds. The appreciation of beauty grows with the experience of living and none of us knows just how much our surroundings influenced the growing mind. To be brought up in Cambridge was itself an advantage: to its beauty of architecture, open spaces, street rivulets and avenues could be joined the fact that it was not so large that one felt hemmed in. The countryside was near at hand with numerous spots by the river or on the chalk hills which seemed designed for picnics and outings. Two uncles had large gardens and in fact the road on which we lived had houses with good gardens. Our house was one of the first to be built along it, and later houses had younger gardens of great variety. One had an array of standard roses, another a border of mixed auriculas; one was annually bedded out with the most splendid begonias and geraniums; there was another where a great yellow lily looked over the fence; most had flowering trees, lilacs and the sort of shrub that could be bought for a shilling or so on the market square. The flowering trees were mostly coloured may, prunus and laburnum; Japanese cherries had not yet 'arrived'. Some had a drive leading in and out; others, not so blessed, had merely a hand-gate. No two were alike . . .

Although my father's garden was of the long and narrow shape associated with smaller houses, it was as intriguing as its dimensions allowed and contained a varied assortment of plants. In fact, looking back on it now, it was obviously very much a product of the time. He was an artist by nature and a draughtsman by training, and made a simple and serviceable design; planted shrubs and fruit trees, built rustic arches and an arbour, revered his lawn (being a golfer) and his hedges. His account book shows that the hedge-clipping shears were purchased for 5s od in 1900; they are still serviceable, likewise his fork. Shortly after this entries are made for two pears for 6s 6d: 'Doyenne du Comice' and 'Williams' Bon Chrêtien'. Other fruit trees amounted to 15s od; they were three apple 'Chivers' Seedling' and one 'Normanton Wonder', one nectarine, five greengage 'Early Transparent' and four gooseberries. He was particularly fond of greengages – and so were we all, in spite of their barren years and copious greenfly. For 13s 3d the garden was stocked with some shrubs, and bulbs amounted to 12s 4d.

The shrubs included, of course, that favourite of the time a 'Spotted Laurel'; the dark green *Euonymus*, which used to suffer badly in cold winters, as did its hardier form with broader leaves edged with yellow; against a sunny wall was trained the white-edged variety and thus it did not suffer. A common green holly and a variegated Hedgehog Holly, a cypress, spring-flowering pink Tamarisk and sweet-smelling 'syringa', Virginian Creeper, a spring-flowering white *Clematis* (*C. montana*) and small summer-flowering purple (*C. viticella*), yellow Winter Jasmine and white Summer Jasmine and a Christmas tree, which was, like the greengages, big enough to climb, completed the woody plants apart from a variegated ivy and an *Ampelopsis* which covered the front of the house. The latter, as is its wont, blocked the drains and had to be removed.

To these were added before long some roses: 'Crimson Rambler' – its only fragrance being that of mildew! – a rose of 1893; 'Alister Stella Gray' a year later in raising and 'Dorothy Perkins' raised in 1901. There was also a very dark red climbing rose whose name was not known. Since I am writing of the years of the First World War, it will be seen that the crimson, yellow and pink ramblers were comparatively 'new' roses, and novelties did not spread quickly in those days. There were also six 'Mrs John Laing' – no doubt cheaper at the half-dozen – and two 'Baroness Rothschild', both older, established varieties, and both silvery pink. 'Mrs John Laing' and 'Alister Stella Gray' remained favourites with us for their fragrance, and they were, in fact, included among the first roses I put into my present garden. Apart from a crimson clove carnation, the rose 'Alister Gray' was a special favourite with my father for a buttonhole, which he wore whenever possible throughout the summer and autumn. Few roses are more fragrant, petite or prolific.

The roses and fruit trees were known by my parents under their proper names; the shrubs were just euonymus, tamarisk, syringa, etc. I think the purchase of these plants all those years ago may be of interest: they were typical of the period before the great awakening to the beauty of flowering shrubs; a few others such as lilac and Snowball Tree, 'Japonica' and *Forsythia* grew in neighbouring gardens.

Snowdrops, daffodils – yellow trumpet 'Princeps' and early and late Pheasant Eye – and 'Glory of the Snow' were joined by many garden toys from my mother's family's large garden at Harston – such as violets and primroses, arabis and aubrieta for edges of borders and the tiny rockery; Lily of the Valley, Solomon's Seal, common ferns, pink and white Japanese Anemones and the native hellebore for shady borders. Elsewhere were orange montbretia, gaillardias, phloxes in several colours, 'Maidenhair Fern' (in reality *Thalictrum adiantifolium*, for long accepted as the ideal foil for cut carnations and sweet peas); and some Michaelmas Daisies including *Aster tradescantii* (of gardens) whose dainty white sprays usually coincided with the early flowers of Winter Jasmine or late chrysanthemums. Irises were in four colours which I can now identify as the old two-toned purple flag, *Iris germanica*; its close relative in dark purple, *I. kochii*; pale yellow *I.* × *flavescens*; and *I. albicans*, the greyish-white whose rhizome is used to produce Orris root. Cornflowers, Shirley Poppies, Larkspur and Nasturtiums were encouraged to seed themselves.

For so small and stereotyped a garden it was a collection of considerable range, I think. From time to time additions were made,

as for instance when a beautiful fern – the Lady Fern, as I now know –
and a Globe Flower were brought back by a cousin from a holiday at
Buttermere, or kind friends produced a root of the Fulvous Day Lily
or the pale lavender *Iris pallida dalmatica* . . .

No attempt was made to grow vegetables, until my brother started
keeping rabbits, which necessitated the growing of kohlrabi. Fortunately for me gardening was not in his blood, and as soon as the
enthusiasm for rabbits was over, I held undisputed sway over the
rubbish area – with no desire to grow anything but flowers.

My father's account book shows regular purchases at the local
nursery, where there were greenhouses full of delectable smells and
pot plants for the house, and where bedding plants could be bought.
Immediately the 'Glory of the Snow' had died down, bedding plants
were put around the euonymus shrubs, usually pansies in mixture,
followed by wallflowers and forget-me-nots in autumn. The pot plants
– mainly beautiful ferns, cyclamens and hyacinths – joined the inevitable palm and green and variegated *Aspidistra* on bamboo stands and
in an old wine-cooler in our windows. But my chief memory of Mr
Hobday's greenhouses is that one day I spied a poor wizened little
pink azalea, thrown away. I asked if I could have it. Planted with
tender care it turned out to be 'Hinomayo' and survived for several
years. Our soil was poor, tacky, and limy and needed all the leafmould
provided by the fruit trees, which was carefully stacked and used; it
was occasionally augmented by bags of well-rotted horse manure from
another uncle whose business entailed keeping the animals for his
removal vans.

I did not for many years realize that there was anything surprising in
this range of garden plants, nor that the garden was more than a patch
of ground for games and 'growing things'. But looked at in the light
of garden history, it carried decided Victorian overtones in its evergreens and its fernery, as well as a grasp of the essentials in plants and
an unobtrusiveness in design, the outcome of the study of design and
art rather than any horticultural prowess from my father.

GRAHAM STUART THOMAS, 1983

It was in a little garden in a rural parish of the county of East Lothian that I began my life amongst flowers. I spent my time after school hours in that garden owned by my grandfather and descended to him from his forefathers. I had to walk a distance of seven miles to get there from my home, for in those days – the penultimate decade of the last century – there were no buses and no bicycles, and the railway was not convenient. Distance, however, mattered nothing, and the effort to get there disappeared altogether in the enthusiasm for the object in view.

The garden was one of three-quarters of an acre, and it was very old. It stood on the site of an ancient monastery which had been founded by the Cistercians as a branch of their abbey at Newbattle, in Mid Lothian, and they always selected such places for the beautiful things of peace. The garden was surrounded with wonderfully massive, moss-grown walls, which the mould of centuries had tinged with that rare legacy of time which cannot be imitated. The old gable of one of the monkish chapels had survived and remained, and was draped in bewitching loveliness with honeysuckle, jasmine and sweet briar, the whole surroundings being incensed with their perfume . . . The soil, like all that had been tended by the monks, was in a high state of fertility, bearing mute testimony to the skill of the hands that had tilled it five hundred years ago . . .

The Roses were all old kinds grown as big bushes sometimes on the walls but generally in the borders, beautiful things filling the garden with their adorable fragrance. There was the old Celestial Rose, delightful in bud and very sweet. Maiden's Blush, delicate blush and laden with its captivating blossoms. Then there was the old double Alba plena, reputed to be Prince Charlie's Rose. The old Cabbage Rose was a favourite and so, too, was its white form. The old Tuscany Rose, at that time the most splendidly coloured of all the dear old garden Roses, was very conspicuous with its almost black velvet, fragrant flowers. We had no Polyanthas in those days, but the little favourite at that time was De Meaux. It was dwarf and pretty. Quaintly beautiful were the old striped Roses such as York and Lancaster, Cottage Maid, and the very striking Tricolor de Flandre, with its white, rose and purple markings. Souvenir de la Malmaison with its ethereal blossoms of soft pink was greatly esteemed, especially so in autumn. So too was the grand old Bourbon, Madame Isaac Pereire, surely one of the sweetest of all Roses, and very captivating with its big mops of vivid carmine flowers. Outstanding amongst the China Roses was the climbing form of Cramoisie Superieure, with its rich, deep crimson blossoms. These attractive flowers were in evidence all through the summer and autumn, and I still regard this Rose as the best of all crimson-flowered climbers. One or two old Tea Roses did well. The kinds I remember were Homere, Maman Cochet, Anna Olivier, and Perle des Jardins. A big bush of the ancient yellow Scottish Rose, *Rosa ochroleuca*, one of the *spinosissimas*, was very lovely. It is possibly the most brilliant golden yellow of all the single Roses. All the Roses were on their own roots – no briar or rugosa suckers to bother with.

There was no bedding out or any attempt at a planned herbaceous border, the flowers grew in delirious promiscuity and seemed to blend in a glorious medley of colour. Snapdragons would rub shoulders with Phloxes, and Batchelor's Buttons with perennial Sunflowers. The old double crimson Sweet William (*Dianthus barbatus magnificus*) vied with the old laced Pinks and the laced silver and gold polyanthuses. The Carnations were the flakes, bizarres and picotees, and outstanding amongst them was Black Diamond. I wish I could get a plant of it now, and also the Double Black Pansy. Both, I am afraid, are lost to cultivation. Double Rockets and Hepaticas did well. Auriculas, the old Dusty Millers, grew abundantly alongside their cousins the Primroses, one of the outstanding of the latter being Madame Pompadour, double, dark, velvety crimson. The old double Crimson Wallflower was a lovely thing. Quite different from the German doubles that were popular in pre-war days. This old double Wallflower was supposed to be lost, but I have come across it again. Tulips, flamed and feathered, were in abundance. Most of the old-time flowers grew in that garden. Lavender, Rosemary, and Artemisia too. What a blending of fragrance, yielding a scent that no French perfume synthesist could rival.

I have seen them, those whom I have named, and as many whom I have forgotten, all thus collected in that old garden. The flowers were drawn up in rows, some according to their kinds, others according to their shapes and shades, while others, as Maeterlinck put it, blended according to the happy chances of the wind and the sun, the most hostile and murderous colours, in order to show that nature acknowledges no dissonance and that all that lives creates its own harmony.

Yes, that old garden taught me many things, and especially that dissidence so far as colour in plants and their planting is concerned, has its virtues as well as its faults. This unexaggerated formalism – so typical of the Elizabethan gardens – enabled one to admire individual plants and their full growth.

GEORGE TAYLOR, 1948

I last had a garden of my own when I was eighteen. Since then my main occupation has been designing gardens for other people. I have worked in England, France, Belgium, Switzerland and in Italy, occasionally in Egypt, once even in Persia, and in the Eastern United States. I have also seen gardens in India, in Ceylon, in Isfahan, in the Lebanon, Scandinavia, Holland and Germany. I have planted window-boxes and cottage gardens, housing schemes for industrial workers, layouts for factories. I have worked for landowners and great industrialists, for corporations and companies, for the very rich and for the poor, for professionals and for amateurs. Through the years this has added up to a wide and special experience. I write 'wide' deliberately. I know nothing whatever of many aspects of gardening and very little of a great many more. But I never saw a garden from which I did not learn something and seldom met a gardener who did not, in one way or another, help me. Perhaps if I had spent these last thirty years making my own garden I would want to share that different experience. As it is, whatever the terms or the place, however different the physical circumstances, I have always tried to shape gardens each as a harmony, linking people to nature, house to landscape, the plant to its soil. This is a difficult standard to achieve and realization has always fallen far short of the concept. At each new attempt, I see that which is superfluous, that is, everything which clutters up my understanding of a problem, must be discarded. Everything which detracts from the idea of a unity must go.

I started to understand something about plants by handling them. It was on one summer holiday when I was perhaps fourteen that, bored with the riding and jumping competitions at a local agricultural show, I wandered off to the flower-tent. There in an atmosphere hot and heavy with the smell of trampled grass, people, animals and flowers, my attention was caught by a tiny plant of *Campanula pulla* with three deep purple bells, huge in comparison with its frail leaves and the minute pot in which it grew. It was mine for a shilling and it opened a new world for me. I had no idea what to do for it nor how to make it flourish in the cold clay of a Lincolnshire garden. So off I went to the public library and within a few days I had found friends and teachers in Reginald Farrer with his *English Rock Garden* and Gertrude Jekyll with *Wall and Water Garden*, two people who had spent a life time with plants and gardens.

All my pocket money went on rock plants. All my holidays were given to my own personal corner of the garden. I would bicycle for miles to get a basket of leaf-soil, I would steal grit, sand or gravel from roadside heaps and I would borrow a horse and cart to collect stones which were hard to come by in our stoneless countryside. My campanula died but meanwhile I had seen a picture of *Primula farinosa*, fallen in love with it and learned that it grew wild in Yorkshire. I had to lure my father, who liked bird-watching, into the Yorkshire dales. There I walked miles questioning every passer-by and after a three weeks' search I eventually found an abandoned quarry starred with the pale mauve treasures that I sought.

When I was a small child there was a market each Friday in the old

Palladian butter market near the Stonebow in Lincoln. The farmers' wives would drive in early in the morning, dressed in their best, with baskets of fresh butter, eggs, chickens, ducks and bunches of freshly picked mint and sage. I used often to be taken there by my grandfather's housekeeper while she made her purchases, and I remember that always, in the spring, there would be bunches of double mauve primroses and of the heavenly-scented *Daphne mezereum*. Later when my passion for gardening developed I wanted these plants but could never find them in our friends' gardens. They seemed to grow only in cottage gardens in hamlets lost among the fields and woods. I gradually came to know the cottagers and their gardens for miles around, for these country folk had a knack with plants. Kitchen windows were full of pots with cascades of *Campanula isophylla*, geraniums, fuchsias and begonias all grown from slips. I would be given cuttings from old-fashioned pinks and roses which were not to be found in any catalogue, and seedlings of plants brought home perhaps by a sailor cousin – here was a whole world of modest flower addicts.

It must have been my father who told me of a certain elderly lady devoted to flowers who lived in a Victorian Gothic house almost in the shadows of the three great towers of Lincoln Minster. One day I knocked on her door. She opened it herself and stood there, tall and gaunt, with wild grey hair and the relics of great good looks, dressed in the fashion of thirty years before. 'Please be careful where you walk,' said the lady – a necessary warning as half the coloured encaustic tiles flooring the dark hallway had been taken up and one had to play hopscotch to avoid a chequer board of Asiatic primula seedlings which grew in the spaces left by the missing tiles. The drawing-room was gardened in another way; ivy had been brought in through holes in the wall to garland windows, walls and ceiling with green. This lady had lived in India where, over many years, she had made lively precise water-colour drawings of flowers, musical instruments, jewels and household objects which filled a whole pile of albums. Outside, in an old sycamore tree, a rickety bamboo ladder led up to a platform among the branches which she called her 'machan', though the neighbour's cat was the only tiger she could stalk. There was a rock garden too, contrived as a home for frogs, lizards and grass snakes. Finding it colourless in the winter, she had imposed on it colonies of brightly coloured toad-stools which she told me she made herself from the lids of boot-polish tins and old toothbrushes. I was always welcome. There were no set meal-times; 'A little food every two hours is better,' she would say, bringing me a plate of pineapple, or custard, or a sandwich.

RUSSELL PAGE, 1962

Most people do not pick their flowering shrubs, but we always did. I can remember the succession of flowering branches, plucked by the adults of the household and arranged by them in a tall gray Chinese jar, in our gold-and-green parlor. My sister and I and our friends had a game we played with the shrubbery. It was called Millinery. All the little girls in the neighborhood would bring to our lawn their broad-brimmed straw school hats, which, because they were Boston girls' hats, had only plain ribbon bands for decoration. Then each of us would trim her straw with blossoms from the shrubs. There was a wide choice of trimmings – forsythia, Japanese crab, Japanese quince, mock orange, flowering almond, lilac, hawthorn, bridal wreath, weigela, deutzia, with its tiny white bells, and, in June, altheas and shrub roses. We were not allowed to pick the rhododendrons or the azaleas, but nothing else was forbidden. When our flowery concoctions were completed, we put them on our heads and proudly paraded into the house to show them off to our elders; it seems to me now that we must have made quite a gay sight. By dusk the trimmings were dead, and the next day we could start all over again.

KATHARINE S. WHITE, 1979

In my garden there are roses with wonderfully evocative names: 'Queen Victoria', 'Giant of Battles', 'Pearl of Gold', 'Rose of the King', 'Baroness Rothschild', 'Marchioness of Lorne'. They read like a roll call of the ancien regime, and in fact these roses are a living link with the past. They are so-called old roses, the types beloved of our great-grandparents. Once the pride of gardeners from New York to San Francisco, they were for generations virtually unobtainable, lost entirely or preserved only in the gardens of a few antiquarians. Today, they are returning, once again filling gardens with their subtle, unfamiliar colors and perfumes. Behind their reappearance lies an extraordinary story, a tale of flowers that have persisted unchanged for centuries and of the unlikely band of experts who united to rescue them from extinction.

My own introduction to old-fashioned roses was accidental, but almost inevitable, since I discovered gardening through the pages of a two-thousand-year-old agricultural treatise. Classics were my first love, most particularly Latin literature and archaeology, but three years of chasing Caesar's ghost through the dusty stacks of the Brown University library left me ready for a change. In the fall of my junior year, I spent a semester in Rome. That city, where modern tenements and shops squat in the shells of ancient monuments, fascinated me. The obtrusion of a broken column or triumphal arch only accentuated the vitality of the streets, the colors, the noise, the smell of the foods and the bite of cold Frascati wine. One evening, a favorite professor and I climbed a hill on the outskirts to watch the sun set over the city. As I drank in the rich, hazy vistas of reds, grays and browns, he turned to me, smiled, and said, 'What a pity we can't dig it all up.'

A few days later in the library, I stumbled over a stray volume of Marcus Porcius Cato's *De Agricultura – On Farming*. Within a few pages, I knew I'd discovered my escape. Nicknamed 'the Censor', Cato was the moral conscience of the second century BC; and in this book he urged his fellow Romans to abandon the baths and banquets and return to a simpler, rural way of life. 'In his early manhood,' the Censor contended, 'the head of the household should be eager to plant his land.' I found that I was, or at least I thought I was after reading Cato's advice on buying a farm and establishing vineyards, orchards and gardens. It was from farmers that the sturdiest men came, or so the old Roman led me to believe. Likewise, it was men engaged in this pursuit who were, according to the same authority, least given to thinking evil thoughts.

I didn't want to doubt the Censor. He was the man, I learned, who succeeded in banning all Greek philosophers and rhetoricians from the city of Rome, a coup that won my jealous admiration, since I was that semester waging a desperate battle with the devious complexities of their prose. Fired by Cato's vision of a return to the soil, I abandoned plans of an academic career and upon my return to Brown struck a deal with the dean for an early graduation so that I could enroll immediately in a horticultural training program at the New York Botanical Garden.

There the gardeners taught me never to call soil 'dirt' (an unforgivable slight to that precious commodity), to swear in Sicilian and to work all day doubled over, because kneeling, even though it might ease my back, would decrease my productivity. They taught me, too – especially the older gardeners, Ralph, Dominic and Patsy – the dignity that comes from a life of collaboration with nature. In 1976, after a two-year apprenticeship, I graduated and had the good fortune to find immediately a job that called on all my skills, academic as well as horticultural. Columbia University commissioned me to restore an old estate it had been given, 126 acres of woods, lawns and gardens along the crest of the Hudson River Palisades.

The landscaping had been done in 1929; it was one of the few great gardens to emerge from the crash and subsequent depression, and it had been built in the grand style. Olmsted Brothers of Brookline, Massachusetts, the successors of Frederick Law Olmsted, the designer of Central Park, had drafted the plans. One hundred and eighty-seven sheets drawn in the most exquisite detail, the blueprints included the name and location of every tree, shrub or flower in the original gardens.

Though I recognized the species names, very few of the cultivars were familiar to me. Of the 27,500 bulbs planted in the fall of 1930, for example, I didn't know a single one. But that didn't surprise me. Fashions in flowers change as regularly as they do in clothes and cars, and nurserymen are quick to dispose of last-year's-model tulips or hyacinths. But roses were, and are, another matter. A bed of roses in full bloom, especially if it is a bit wild and overgrown, gratifies in me a repressed yearning for colorful, sensual disorder. Roses I had grown by the hundreds, and I prided myself on my ability to maintain them in an apparent state of romantic abandon while secretly cultivating them in perfect health . . .

Like most gardeners, I have an abiding mistrust of progress. Synthetic fertilizers, turbo-charged tools and genetic engineering may fill agriculturists and extension agents with visions of utopia, but they are far less attractive to the practitioners of a craft that has not changed in its essentials since Pliny the Younger's slaves spelled out their master's name in boxwood hedges.

THOMAS CHRISTOPHER, 1989

G iven my head I should be a Grand Turk in compostery: with agents scouring the countryside for the best peat, sand, alluvial mould, and yellow loam; with a large building (probably the size of an aeroplane hangar) devoted exclusively to the storing and manufacture of compost, with all the constituents in rat-proof, damp-proof containers, each labelled with name and year like bins of vintage claret; and with benches raised three feet from the floor on which to mix up the ingredients until, as the experts say, they are 'intimately blended'.

But the setting up of such a manufactory would hardly be an economy measure. It was a question of removing my head from the clouds, collecting my own materials, using the garage as a store, and mixing the composts myself: for sowing, two parts of loam to one part of peat and one part of sand; and, for growing, seven parts loam, three parts peat, and two parts sand, plus a handful of lime and John Innes Base to each bucketful.

Loam and sand were to hand, but I had to go a fair distance for supplies of peat, or rather leaf mould, and one embarrassing encounter put me off it. It happened, I recall, in the second year of my glass-gardening experiments under the general supervision of Villikins.

A friend and I went to a woodland in a van to collect a load of fine leaf mould. We arrived, unloaded two shovels and a black cabin trunk in which we intended to carry the leaf mould to the van, removed our jackets and set to work. A piercing and extremely alarming cry from the depths of the wood told us we were not alone. It appeared there was a permanent resident there in a bracken- and bramble-covered caravan. It appeared also that not unreasonably she thought (at worst) we had a chopped-up corpse to dispose of in our black cabin trunk, or (at best) that we were stealing her property. She would not reveal herself but shrieked through a gap in the brambles. When we drew near to reason with her she shrieked the louder. She would not acknowledge we stood on common land. She quoted the Ten Commandments at length and at full volume. In such a case as this it is

wiser not to hang about. Like Burke and Hare we stuffed the shovels
and the trunk into the van and drove like the wind, and even above
the noise of the engine we heard for a long time the caravanner's
imprecations – 'Thou shalt not steal!' she was howling. 'Thou shalt not
steal!'

TYLER WHITTLE, 1969

Returning on a summer's afternoon from a parochial walk, I in-
ferred from wheel-tracks on my carriage-drive that callers had
been and gone. I expected to find cards in the hall, and I saw that the
horses had kindly left theirs on the gravel. At that moment one of
those

> Grim spirits in the air,
> Who grin to see us mortals grieve,
> And dance at our despair,

fiendishly suggested to my mind an economical desire to utilize the
souvenir before me. I looked around and listened; no sight, no sound,
of humanity. I fetched the largest fire-shovel I could find, and was
carrying it bountifully laden through an archway cut in a high hedge of
yews, and towards a favourite tree of 'Charles Lefebvre', when I suddenly confronted three ladies, 'who had sent round the carriage, hearing that I should soon be at home, and were admiring my beautiful
Roses'. It may be said with the strictest regard to veracity, that they saw
nothing that day which they admired, in the primary meaning of the
word, so much as myself and fire-shovel; and I am equally sure that no
Rose in my garden had a redder complexion than my own.

S. REYNOLDS HOLE, 1885

G ardener. – Thomas Valentine, Bred under the ablest master in
Ireland, who for some years after his apprenticeship conducted
the Gardening Business for the Right Hon. the Earl of Belvedere, a
Nobleman remarkable for elegant Taste, extensive Gardens and Plantations, the Major part of which were made immediately under said
Gardener's Direction, during his Services with him; and has been
employed by several of the Nobility and Gentry, to lay out their Gardens and Improvements. He also surveys Land, makes Copies and
traces Maps, draws Designs for Gardens, Plantations, Stoves, green
Houses, forcing Frames, etc. etc. He is willing to attend any Gentleman's Gardens within ten or twelve Miles of this City [New York], a
Day or two in the Week, and give such Directions as are necessary for
completing and keeping the same in proper Order . . .

New York Gazette and the Weekly Mercury, 1768

H onestie in a Gardner, will grace your Garden, and all your house
and help to stay unbrideled Serving-men, giving offence to
none, not calling your name into question by dishonest acts, nor infect-
ing your family by evill counsell or example. For there is no plague so
infectious as Popery and Knavery, hee will not purloin your profite,
nor hinder your pleasures.

Concerning his skill, hee must not be a Scholist, to make shew of, or
take in hand that, which he cannot performe, especially in so weighty
a thing as an Orchard: then the which, there can be no humane
thing more excellent, eyther for pleasure or profit. And what an
hinderance shall it bee, not only to the owner but to the common
good, that the unspeakable benefits of many hundred yeares, shall

be lost, by th' audacious attempt of an unskilful Arborist.

The Gardener had not need be an idle or lazy Lubber, for so your Orchard, being a matter of such moment, will not prosper, there will ever be something to do. Weeds are alwaies growing, the great Mother of all living Creatures, the Earth, is full of Seed in her Bowels, and any stirring gives them heat of Sun, and being laid near day, they grow: Moles work daily, though not alwaies alike: Winter Herbs at all times will grow (except in extream Frost). In Winter your Trees and Herbs would be lightened of Snow, and your Allies cleansed: drifts of Snow will set Deer, Hares and Conies, and other noysome Beasts, over your Walls and Hedges into your Orchard. When Summer cloaths your Borders with Green and speckled colours, your Gardener must dress his hedges, and antick works; watch his Bees, and hive them: Distil his Roses and other Herbs. Now begin Summer Fruits to ripen, and crave your hand to pull them. If he have a Garden (as he must needs) to keep, you must needs allow him good help, to end his labours which are endless; for no one man is sufficient for these things.

Such a gardner as will conscionbly, quietly and patiently travell in your Orchard God shall crowne the labors of his hands with joyfulnesse, and make the cloudes droppe fatnesse upon your Trees, hee will provoke your love, and earne his Wages, and fees belonging to his place: The house being served, fallen fruict, superfluity of hearbes, and floures, seedes, graffes [grafts], sets and besides all other of that Fruit which your bountifull hand shall award him withall: will much augment his Wages, and the profite of your bees will paye you backe againe.

If you bee not able, nor willing to hyre a gardner, keepe your profites to your selfe, but then you must take all the paines.

WILLIAM LAWSON, 1618

The first time that I saw him, I fancy Robert was pretty old already: he had certainly begun to use his years as a stalking horse. Latterly he was beyond all the impudencies of logic, considering a reference to the parish register worth all the reasons in the world. 'I am old and well stricken in years,' he was wont to say; and I never found any one bold enough to answer the argument. Apart from this vantage that he kept over all who were not yet octogenarian, he had some other drawbacks as a gardener. He shrank the very place he cultivated. The dignity and reduced gentility of his appearance made the small garden cut a sorry figure. He was full of tales of greater situations in his younger days. He spoke of castles and parks with a humbling familiarity. He told of places where under-gardeners had trembled at his looks, where there were meres and swanneries, labyrinths of walk and wildernesses of sad shrubbery in his control, till you could not help feeling that it was condescension on his part to dress your humbler garden plots. You were thrown at once into an invidious position. You felt that you were profiting by the needs of dignity, and that his poverty and not his will consented to your vulgar rule. Involuntarily you compared yourself with the swineherd that made Alfred watch his cakes, or some bloated citizen who may have given his sons and his condescension to the fallen Dionysius. Nor were the disagreeables purely fanciful and meta-physical, for the sway that he exercised over your feelings he extended to your garden, and, through the garden, to your diet. He would trim a hedge, throw away a favourite plant, or fill the most favoured and fertile section of the garden with a vegetable that none of us could eat, in supreme contempt for our opinion. If you asked him to send you in one of your own artichokes, 'That I wull, mem,' he would say, 'with pleasure, for it is mair blessed to give than to receive.' Ay, and even when, by extra twisting of the screw, we prevailed on him to prefer our commands to his own inclination, and he went away, stately and sad, professing that 'our wull was his pleasure', but yet reminding us that he would do it 'with feelin's', – even then, I say, the triumphant master felt humbled in his triumph, felt that he ruled on sufferance only, that he was taking a mean advantage of the other's low estate, and that the whole scene had been one of those 'slights that patient merit of the unworthy takes'.

In flowers his taste was old-fashioned and catholic; affecting sunflowers and dahlias, wallflowers and roses, and holding in supreme aversion whatsoever was fantastic, new-fashioned or wild. There was one exception to this sweeping ban. Foxgloves, though undoubtedly guilty on the last count, he not only spared, but loved; and when the shrubbery was being thinned, he stayed his hand and dexterously manipulated his bill in order to save every stately stem. In boyhood, as he told me once, speaking in that tone that only actors and the old-fashioned common folk can use nowadays, his heart grew 'proud' within him when he came on a burn-course among the braes of Manor that shone purple with their graceful trophies; and not all his apprenticeship and practice for so many years of precise gardening had banished these boyish recollections from his heart. Indeed, he was a man keenly alive to the beauty of all that was bygone. He abounded in old stories of his boyhood, and kept pious account of all his former pleasures; and when he went (on a holiday) to visit one of the fabled great places of the earth where he had served before, he came back full of little pre-Raphaelite reminiscences that showed real passion for the past, such as might have shaken hands with Hazlitt or Jean-Jacques.

But however his sympathy with his old feelings might affect his liking for the foxgloves, the very truth was that he scorned all flowers together. They were but garnishings, childish toys, trifling ornaments for ladies' chimney-shelves. It was towards his cauliflowers and peas and cabbage that his heart grew warm. His preference for the more useful growths was such that cabbages were found invading the flower-pots, and an outpost of savoys was once discovered in the centre of the lawn. He would prelect over some thriving plant with wonderful enthusiasm, piling reminiscence on reminiscence of former and perhaps yet finer specimens. Yet even then he did not let the credit leave himself. He had, indeed, raised 'finer o' them', but it seemed that no one else had been favoured with a like success. All other gardeners, in fact, were mere foils to his own superior attainments; and he would recount, with perfect soberness of voice and visage, how so and so had wondered, and such another could scarcely give credit to his eyes. Nor was it with his rivals only that he parted praise and blame. If you remarked how well a plant was looking, he would gravely touch his hat and thank you with solemn unction; all credit in the matter falling to him. If, on the other hand, you called his attention to some back-going vegetable, he would quote Scripture: 'Paul may plant and Apollos may water', all blame being left to Providence, on the score of deficient rain or untimely frosts.

There was one thing in the garden that shared his preference with his favourite cabbages and rhubarb, and that other was the beehive. Their sound, their industry, perhaps their sweet product also, had taken hold of his imagination and heart, whether by way of memory or no I cannot say, although perhaps the bees too were linked to him by some recollection of Manor braes and his country childhood . . .

I could go on for ever chronicling his golden sayings or telling of his innocent and living piety. I had meant to tell of his cottage, with the German pipe hung reverently above the fire, and the shell box that he had made for his son, and of which he would say pathetically: 'He was real pleased wi' it at first, but I think he's got a kind o' tired o' it now' – the son being then a man of about forty. But I will let all these pass. ' 'Tis more significant: he's dead.' The earth, that he had digged so much in his life, was dug out by another for himself; and the flowers that he had tended drew their life still from him, but in a new and nearer way. A bird flew about the open grave, as if it too wished to honour the obsequies of one who had so often quoted Scripture in favour of its kind: 'Are not two sparrows sold for one farthing, and yet not one of them falleth to the ground.'

Yes, he is dead. But the kings did not rise in the place of death to greet him 'with taunting proverbs' as they rose to greet the haughty Babylonian; for in his life he was lowly, and a peacemaker and a servant of God.

ROBERT LOUIS STEVENSON, 1904

It would be an anomaly to find a student of nature addicted to the vices that cast so many dark shadows on our social life; nor do I remember among the sad annals of criminal history, one instance of a naturalist who became a criminal, or of a single gardener who has been hanged.

SHIRLEY HIBBERD, 1856

Since I went away there has been a change in the garden, Charles having elected to transfer himself to the stable, preferring, I think, the care of ponies to that of flowers. In his place Mason's son has arrived, a boy of nearly fifteen. In watching this little fellow at work I have come to the conclusion that a taste for gardening is hereditary. His active experience in that line has been limited to six months in a neighbouring establishment, yet to all appearance it might have been six years. Already I see him engaged upon quite responsible work, such as the clearing and cleaning of Asparagus beds, or the cutting down and removal of herbaceous things, and hear him reminding his father that it is time to close the ventilators in the houses. Decidedly, like the poet, the gardener must be born, and not made.

H. RIDER HAGGARD, 1905

Gardeners may scold as long and vehemently as they please, and law-makers may enact as long as they please, mankind will never look upon taking fruit in an orchard, or a garden, as *felony*, nor even as a *serious trespass*. Besides, there are such things as *boys*, and every considerate man will recollect that he himself was once a boy. So that, if you have a mind to have for your own exclusive use what you grow in your garden, you must do one of two things: resort to terrors and punishments, that will make you detested by your neighbours, or provide an insurmountable fence. This prevents *temptation*, in all cases dangerous, and particularly in that of forbidden fruit. Resolve, therefore, to share the produce of your garden with the boys of the whole neighbourhood; or, to keep it for your own use by a fence that they cannot get through, over, or under. Six feet is no great height; but in the way of *fence*, four feet of good thorn-hedge will keep the boldest boy from trees loaded with fine ripe peaches; and, if it will do *that*, nothing further need be said in its praise! The height is nothing; but unless the assailant have wings, he must be content with feasting his eyes; for, if he attempt to *climb*, he receives the penalty upon the spot; and if he retreats, as the fox did from the grapes, he gets pain of body in addition to that of a disappointed longing. I really (recollecting former times) feel some remorse in thus plotting against the poor fellows; but the worst of it is, they will not be content with fair play: they will have the *earliest* in the season, and the *best*, as long as the season lasts; and, therefore, I must, however reluctantly, shut them out altogether.

WILLIAM COBBETT, 1829

THE DIEHARDS

We go, in winter's biting wind,
On many a short-lived winter day,
With aching back but willing mind
To dig and double-dig the clay.

All in November's soaking mist
We stand and prune the naked tree,
While all our love and interest
Seem quenched in blue-nosed misery.

We go in withering July
To ply the hard incessant hoe;
Panting beneath the brazen sky
We sweat and grumble, but we go.

We go to plead with grudging men,
And think it is a bit of luck
When we can wangle now and then
A load or two of farmyard muck.

What do we look for as reward?
Some little sounds, and scents, and scenes:
A small hand darting strawberry-ward,
A woman's apron full of greens.

A busy neighbour, forced to stay
By sight and smell of wallflower-bed;
The plum-trees on an autumn day,
Yellow, and violet, and red.

Tired people sitting on the grass,
Lulled by the bee, drugged by the rose,
While all the little winds that pass
Tell them the honeysuckle blows.

The sense that we have brought to birth
Out of the cold and heavy soil,
These blessed fruits and flowers of earth
Is large reward for all our toil.

RUTH PITTER, 1941

I will not betray to you how gardeners recognize one another, whether by smell, or some password, or secret sign; but it is a fact that they recognize one another at first sight, whether in the gangways of the theatre, or at a tea, or in a dentist's waiting-room; in the first phrases which they utter they exchange views on the weather ('No, sir, I never remember such a spring'), then they pass to the question of humidity, to dahlias, artificial manures, to a Dutch lily ('damned thing; what's its name, well, never mind, I will give you a bulb'), to strawberries, American catalogues, damage from last winter, to aphis, asters, and other such themes. It is only an illusion that they are two men in dress suits in the gangway of the theatre; in deeper and actual reality they are two gardeners with a spade and a watering-can.

KAREL ČAPEK, 1931

The care of plants, such as needed peculiar care or skill to rear them, was the female province. Every one in town or country had a garden. Into this garden no foot of man intruded after it was dug in the Spring. I think I see yet what I so often beheld – a respectable mistress of a family going out to her garden, on an April morning, with her great calash, her little painted basket of seeds, and her rake over her shoulder, going to her gardens of labours. A woman in very easy circumstances and abundantly gentle in form and manners would sow and plant and rake incessantly.

ANNE GRANT, 1808

I went to Lordship's when I was fourteen and stayed for fourteen years. There were seven gardeners and goodness knows how many servants in the house. It was a frightening experience for a boy. Lord and Ladyship were very, very Victorian and very domineering. It was 'swing your arms' every time they saw us. Ladyship would appear suddenly from nowhere when one of us boys were walking off to fetch something. 'Swing your arms!' she would shout. We wore green baize aprons and collars and ties, no matter how hot it was, and whatever we had to do had to be done on the dot. Nobody was allowed to smoke. A gardener was immediately sacked if he was caught smoking, no matter how long he had worked there.

We must never be seen from the house; it was forbidden. And if people were sitting on the terrace or on the lawn, and you had a great barrow-load of weeds, you might have to push it as much as a mile to keep out of view. If you were seen you were always told about it and warned, and as you walked away Ladyship would call after you, 'Swing your arms!' It was terrible. You felt like somebody with a disease.

The boy under-gardeners had to help arrange the flowers in the house. These were done every day. We had to creep in early in the morning before breakfast and replace great banks of flowers in the main rooms. Lordship and Ladyship must never hear or see you doing it; fresh flowers had to just be there, that was all there was to it. There was never a dead flower. It was as if flowers, for them, lived for ever. It was part of the magic in their lives. But the arrangements were how they wanted them and if one of the gardeners had used his imagination, Ladyship noticed at once and soon put a stop to it! The guests always complimented her on the flowers and she always accepted the praise as though she had grown, picked and arranged them herself. It was logical because servants were just part of the machinery of the big house and people don't thank machines, they just keep them trim and working. Or that's how I look at it.

As the years went by, we young men found ourselves being able to talk to Lordship and Ladyship. 'Never speak to them – not one word and no matter how urgent – until they speak to you,' the head-gardener told me on my first day. Ladyship drove about the grounds in a motor-chair and would have run us over rather than have to say, 'get out the way'. We must never look at her and she never looked at us . . . We were just there because we were necessary, like water from the tap. We had to listen for voices. If we heard them in a certain walk, we had to make a detour, if not it was, 'But why weren't you listening?' and 'Be alert, boy!' and, when you had been dismissed, 'Swing your arms!'

The garden was huge. The pleasure grounds alone, and not including the park, covered seven acres. The kind of gardening we did there is not seen nowadays. It was a perfect art. Topiary, there was a lot of that. It was a very responsible job. You had only to make one bad clip and a pheasant became a duck. The gardeners usually made up these creatures themselves. We were tempted to cut out something terrible sometimes, so that it grew and grew . . . but of course we never did. Even when we went on to mechanical hedge-trimmers we still kept on topiary. There was a great pride in it, and in hedge-cutting of every sort. It was the hedge which set the garden off and all the big houses competed with each other. Fences were marvellous things, too; there were more than two miles of them round Lordship's and not a pale which wasn't exact. The hedges had tops like billiard-tables. It was get down and have a look, and stand back and have a look. No hedge was left until it was marvellous. There were so many things which really had no need to be done but which we did out of a kind of obstinate pleasure. The asparagus beds in winter were an example. We'd spend hours getting the sides of the clamps absolutely flat and absolutely at a 45° angle, although an ordinary heap of earth would have done just as well.

None of the village people were allowed into the garden. Definitely not. Trades-people came to their door and never saw the main gardens. Work in front of the house had to be done secretly. About seven in the morning we would tiptoe about the terrace, sweeping the leaves, tying things up, never making a sound, so that nobody in the bedrooms could hear the work being done. This is what luxury means – perfect consideration. We gave, they took. It was the complete arrangement . . .

Of course, they spent a terrific amount of money on the house and garden. It was the machinery they had to have in order to live. So they kept it going, as you might say. A bad servant was just a bad part and was exchanged for a good part as soon as possible. I thought of this when I was doing my National Service as a fitter in the R.A.C. [Royal Army Corps]. It made sense. Yet I got so that I didn't know quite what to think about it all. It was obviously wrong, yet because Lordship and Ladyship were old and had never known any other kind of life, I suppose I felt sorry for them. I always had to give more than was necessary. I couldn't resist it. It was exciting somehow. But when I got home I would be angry with myself. The butler would sometimes come to the pub and imitate them. Laugh – you should have heard us! But I would feel strange inside, pitying and hating at the same time. His favourite joke was:

Ladyship: 'Shall we ask the So-and-Sos to luncheon, Bertie?'
Silence, then, 'Can they play bridge? Will they like my garden?'

Ladyship: 'No, I don't think so.'

Lordship: 'Then don't have 'em.'

Lordship was a friend of King George V. He was a terribly nice man – a real gentleman. A lot of royalty came down from time to time and Lordship and Ladyship were sometimes at Sandringham. The Queen (Queen Elizabeth the Queen Mother) came. She treated us very well and loved the garden. She would tell us boys what they ate for luncheon and then we'd all laugh. The Princess Royal was just the same – easy. But Members of Parliament always imitated Lordship and Ladyship and treated us like fittings. I was amazed by the Royalty. I imagined a bigger kind of Ladyship, but definitely not . . .

I had a great training as a gardener and acquired all my knowledge completely free. Although I was often horrified by the way we were all treated, I know I got a terrific amount out of it. It is a gardening background which few people now have, and scarcely anybody of my age. In a great garden you grow from the seed and then you see the plant growing where it will always grow, but in a nursery garden it is just produce and sell, produce and sell. Nothing remains. A private gardener like myself would never get on in nursery work because I have had the fine art of tidiness drummed into me. I work privately and could have a choice of twelve or fifteen jobs, all with houses. There is no kind of gardening I can't do. I am not boasting, it is a fact . . .

There aren't many gardeners of my calibre left. I am a young man who has got caught in the old ways. I am thirty-nine and I am a Victorian gardener, and this is why the world is strange to me.

'CHRISTOPHER FALCONER', 1969

The sixteenth century, which saw the English garden formulated, was a time for grand enterprises; indeed, to this period is ascribed the making of England. These gardens, then, are the handiwork of the makers of England, and should bear the marks of heroes. They are relics of the men and women who made our land both fine and famous in the days of the Tudors; they represent the mellow fruit of the leisure, the poetic reverie, the patient craft of men versed in great affairs – big men, who thought and did big things – men of splendid genius and stately notions – past-masters of the art of life who would drink life to the lees.

As gardeners, these old statesmen were no dabblers. They had the good fortune to live in a current of ideas of formal device that touched art at all points and was well calculated to assist the creative faculty in design of all kinds. They lived before the art of bad gardening had been invented; before pretty thoughts had palled the taste, before gardening had learnt routine; while Nature smiled a virgin smile and had a sense of unsolved mystery. More than this, gardencraft was then no mere craze or passing freak of fashion, but a serious item in the round of home-life; – gardening was a thing to be done as well as it could be done. Design was fresh and open to individual treatment – men needed an outlet for their love of, their elation at, the sight of beautiful things, and behind them lay the background of far-reaching traditions to encourage, inspire, protect experiment with the friendly shadow of authority.

JOHN D. SEDDING, 1890

When we first became friends, I suggested to Micah that he should help me regularly in the garden. He had plenty of time, and I felt I could just about spare enough money to pay someone to put in a few hours regularly on the harder work.

He refused at once.

'I'll give you a hand,' he said. 'I'll give you a hand any time you want a bit of help. But not regular, and not for pay.'

I could not see that at all. If he helped me, I ought to pay him.

He shook his head at me, grinning as at some secret thought.

'If I was to be your gardener,' he explained, 'before long it would be my garden, not yours. You'd be thinking you'd better not do this and better not touch that. I'd be having the sack in no time and the both of us middling cross with each other.'

I protested that this was nonsense, but perhaps not with as much conviction as I might have shown. Dimly, I could see what he meant.

'You've got a garden one man can manage – with a bit of help from friends,' he said. 'The sensible thing is to manage it. If you had me regular, you'd be making the one really bad mistake any man can make in a garden.'

'And that is?'

'To have a gardener who knows more about gardening than he does himself.'

He saw my puzzled look.

'Well, that's the way you put it yourself, don't you? You always tell me I know more about gardening than you do.'

He did. He was not boasting. It was a simple fact.

'If you ever take on a gardener, you want to pick one who knows *less* than you do. A willing one, mind. One with heart and soul in the job, but who knows you're boss and knows you know a lot more than he does. Then he comes and asks you what to do and how to do it – and you get things done *your* way.

'Now, you have a chap who knows a lot more than you, and knows he knows a lot more than you do, and he won't even bother to ask for advice or how to do things. He'll do them – *his* way. And you'll have to put up with it. If you start telling him he's wrong, or you want things done different, he can tell you to mind your own business. No, if you want a gardener, you pick one ignorant and willing.'

'I'd better advertise for one,' I said. '"Wanted, a gardener. Must know absolutely nothing about gardening."'

'You could do worse,' said Micah dryly. 'Of course, people who *want* their gardening all done for them need a properly qualified man. But you're one of those interfering chaps. You couldn't leave your garden alone. You'd never do to have a gardener. There'd be fights in no time.'

So Micah is not my gardener. He does help me quite a lot. But I am still master of my own bit of ground.

Later on, thinking over what he had said, I remembered hearing someone say how he had once gone round a famous garden with its owner. On a sunny south wall there were some ripe Peaches, and the man asked my friend would he like one? My friend said he would.

The owner of the garden stepped across the bed between the path and the wall. He had actually stretched out his hand to pluck a Peach, when he thought better of it.

He stepped back carefully, and as he retreated he smoothed away every footprint with the palm of his hand so that it could not be seen that anyone had walked there.

He looked shamefaced as he dusted his hands and grinned apologetically at my friend.

'MacTavish, you know. He wouldn't like it.'

Remembering that, I began to see Micah was right.

H. L. V. FLETCHER, 1948

The distinctive term of 'florists' flowers' represents now, with a larger meaning than it did a quarter of a century ago, a group of subjects, some of which have for many years past been taken in hand by persons specially interested in them; and cultivated, as well as improved, with great care, mainly for the exhibition table. The annals of floriculture are prolific of records showing how workers in various positions in society, but mainly in the humbler walks of life, have taken up one, two, or more subjects (such as the Auricula, Carnation, Pink, and Tulip, among others), and cultivating them with the greatest care, saved seed and raised seedlings; selecting with intelligence from these only such as were manifest improvements upon the varieties known to them, and rejecting all others as inferior types unworthy their attention . . .

One result of this floricultural enterprise was the establishment of a large number of small exhibitions of flowers in many parts of the country, where cultivators of certain subjects could meet, put their specimens into comparison, and receive premiums for the best. These shows sprang up with marvellous rapidity in and around London, and especially in Lancashire, Yorkshire, and other of the Midland and Northern Counties. The Flemish and French weavers, who many years ago were driven from their own country by religious persecution, brought with them their Auriculas and other flowers. Settling in London, Lancashire, and elsewhere, they cultivated them with assiduity and success; and intermarrying with those among whom they came to reside, spread abroad a love for their flowers, and a desire to cultivate them.

RICHARD DEAN, *c.* 1890

Because I have a taste for anything that is quaint and curious, and frequently chance on strange and out-of-the-way people, I count myself a happy man. I once encountered just such a character in an English country lane; a cyclist who was attending to a puncture, although when I first saw him he was resting and standing with his hand on the upturned bicycle as though it was a big-game trophy. It was a fascinating picture, especially as he was quite the most oval man I have ever met and dressed so gaily that from a distance he resembled a decorated Easter egg. I engaged him in conversation – which was desultory and rather a one-sided affair until I remarked on the splendour of the tall Hawthorn hedge which was shading us from the heat of the sun. This was like switching on a gramophone. Thereafter he talked hedges with great enthusiasm and I was enchanted – but somewhat startled when he stopped abruptly, looked up and down the road, as though to make sure he was not being overheard, and whispered secretly: 'I have a weakness for topiary.'

It was such an unexpected and sudden disclosure that I wondered if I could have misheard him, and he had, in fact, confessed to some dark and horrid vice. Apparently not, for in the same hushed voice he made a speech about topiary, and finished by asking me to visit him and inspect his work. I considered it was far too good an opportunity to be missed and, although I was staying with friends, on the next day I excused myself for the whole of the morning and drove over to see the egg-shaped topiarist.

He was a widower who evidently cared nothing for the inside of his house but cared passionately for the garden. This was as neat as a pin and surrounded by a large Yew hedge. In fact he had bought the place – a poky, dark and airless Victorian lodge without running water or electricity – solely because he wanted to attack the hedge with shears, pruning knives and scissors. He had been tinkering with it ever since, but never with a great deal of success, because it was too old and well-established. He showed me his best piece of hedge carving: 'This,' he declared, 'was intended originally to be Field-Marshal Montgomery, but, at a critical stage, his beret got out of control, and I had to turn him into a turkey.' It was a good turkey: thoroughly wattled, beautifully feathered, and as arrogant as a real stag turkey. But his finest carvings were in Privet and Box and Ilex . . . a fish standing on its tail; a question mark; a chesspiece in fine proportions; a man-o'-war's cannon; odd birds and beasts including a Pelican in her Piety; and something which I thought was a chemist's carboy but which, declared the topiarist, was the foundations of a Red Indian's head in variegated Holly.

'You have to plan,' he said. 'Plan in minute detail. See it shaping in your mind's eye. Then snip with care.'

Up to that time I had regarded this ancient garden-craft as too Tudor and too bizarre a decoration for serious consideration, but my new friend made me see it through other spectacles. He took me from piece to piece, demonstrating this and that, explaining that topiary was so especially fascinating because it is based on clairvoyance as well as draughtsmanship and sylviculture. And his zeal was infectious. Since then I have been examining other hedge and shrub carvings and deriving a great deal of pleasure from those which are well contrived and suited to their environment. I have even gone so far as to cut one shape in my Norfolk garden. It is well out of the way, very small, and very simple – a ball of thorn on a long, thin stem; and if anyone inquires what it is supposed to represent I have no hesitation in investing it with dignity. This, I say, is the battle mace of a mediaeval bishop who, barred by his consecration from shedding blood with pike or sword, was still permitted to bash his opponent's skull to pulp whenever the occasion arose. I make no doubt that this exalted explanation, as well as the shape itself, would have the warm approval of my friend the topiarist, but I never had an opportunity to tell him about it or pay a second visit to his garden of curiosities. Before I could return to that part of England he died, and it was painful to learn that his house had been sold to gardeners who disliked topiary and were devoted to wriggly-edged borders of bedding plants; and they had cut his fashioned hedge and other carvings to the ground, clawed out the roots, and burnt them up. Nevertheless, he was an irrepressible enthusiast for his art and, as Paradise is said to be a garden, I am hopeful that the kindly egg-shaped topiarist has been given permission and sharp shears to sculpt the evergreens of Heaven.

TYLER WHITTLE, 1965

M. Emanuel had a taste for gardening; he liked to tend and foster plants. I used to think that working amongst shrubs with a spade or watering-pot soothed his nerves; it was a recreation to which he often had recourse; and now he looked to the orange trees, the geraniums, the gorgeous cactuses, and revived them all with the refreshment their drought needed. His lips meantime sustained his precious cigar, that (for him) first necessary and prime luxury of life; its blue wreaths curled prettily enough amongst the flowers, and in the evening light.

CHARLOTTE BRONTË, 1853

Some joy shall also come of the identity of the gardener with his creation. He is at home here. He is intimate with the various growths. He carries in his head an infinity of details touching the welfare of the garden's contents. He participates in the life of his plants, and is familiar with all their humours; like a good host, he has his eye on all his company. He has fine schemes for the future of the place. The very success of the garden reflects upon its master, and advertises the perfect understanding that exists between the artist and his materials. The sense of ownership and responsibility brings him satisfaction, of a cheaper sort. His the hand that holds the wand to the garden's magic; his the initiating thought, the stamp of taste, the style that gives it circumstance. Let but his hand be withdrawn a space, and, at this signal, the gipsy horde of weeds and briars – that even now peer over the fence, and cast clandestine seeds abroad with every favouring gust of wind – would at once take leave to pitch their tents within the garden's zone, would strip the place of art-conventions, and hurry it back to its primal state of unkempt wildness.

JOHN D. SEDDING, 1890

The gardener here [Stroud House, near Haslemere], Arthur Morrey, is a most industrious and valuable man, and every thing under his charge does him great credit. He also has two boys and four girls, healthy children, to each of whom we have sent a school-book, and, to the father, a pair of French *sabots* for putting over his shoes in the pruning season. We would strongly recommend these *sabots* (wooden shoes) to all journeymen gardeners, as most valuable for keeping their feet dry and warm while standing on wet ground pruning trees in the winter or spring season. They may be had through any London or Edinburgh nurseryman, who may easily procure them from any sea-port on the Continent, and they are very cheap and durable. Indeed, we are of the opinion that every head-gardener ought to keep a stock of them for the use of the men under his care, in the same way as he keeps spades, rakes, and other tools. Nurserymen and gentlemen's gardeners find that it pays to warm, by a flue or a steam-pipe, the back sheds in which their men work in the winter time. Why should it not, also, pay to keep their workmen's feet dry and warm when they are working in the open air at that season? To begin the thing, we hereby offer a copy of our *Hortus Britannicus* to the first head-gardener in England who shall, with the consent of his employer, procure 20 pairs of *sabots* from a London nurseryman, for the use of his men; the like stimulus to the first gardener in Scotland, the *sabots* being procured from an Edinburgh nurseryman; and the like for Ireland, the *sabots* being procured from a Dublin nurseryman. We are desirous that the *sabots* should be procured from nurserymen, in order that these may get into the way of keeping a stock of them; and we shall be glad to know the nurserymen's names, that we may publish them for the benefit of gardeners generally.

JOHN CLAUDIUS LOUDON, 1829

Collected the following plants and obtained seeds of several very important plants already collected: among them *Nicotiana quadrivulvis*, correctly supposed by Nuttall to exist on the Columbia; whether its original habitat is here in the Rocky Mountains, or on the Missouri, I am unable to say, but am inclined to think it must be in the mountains. I am informed by the hunters it is more abundant towards them and particularly so among the Snake Indians, who frequently visit the Indians inhabiting the head-waters of the Missouri by whom it might be carried in both directions. I have seen only one plant before, in the hand of an Indian two months since at the Great Falls of the Columbia, and although I offered him 2 ozs. of manufactured tobacco he would on no consideration part with it. The natives cultivate it here, and although I made diligent search for it, it never came under my notice until now. They do not cultivate it near camps or lodges, lest it should be taken for use before maturity. An open place in the wood is chosen where there is dead wood, which they burn, and sow the seed in the ashes. Fortunately I met with one of the little plantations and supplied myself with seeds and specimens without delay. On my way home I met the owner, who, seeing it under my arm, appeared to be much displeased; but by presenting him with two finger-lengths of tobacco from Europe his wrath was appeased and we became good friends. He then gave me the above description of cultivating it. He told me that wood ashes made it grow very large. I was much pleased with the idea of using wood ashes. Thus we see that even the savages on the Columbia know the good effects produced on vegetation by the use of carbon. His knowledge of plants and their uses gained him another finger-length. When we smoked we were all in all.

Returned on the 30th of August. From that time till Thursday, September 1st, employed drying, arranging, putting up seeds, and making up my notes. Early on Thursday went on a journey to the Grand Rapids to collect seeds of several plants seen in flower in June and July. Went up in a canoe accompanied by one Canadian and a Chief (called Chumtalia) of the tribe inhabiting the north banks of the river at the Rapids. I arrived on the evening of the second day and pitched my tent a short distance from the village. I caused my Canadian to drench the ground well with water to prevent me from being annoyed with fleas, although I was not altogether exempt from them, yet it had a good effect. I found my Indian friend during my stay very attentive and I received no harm or insult. He accompanied me on some of my journeys. (They were only a few years since very hostile. The Company's boats were frequently pillaged by them and some of their people killed.) My visit was the first ever made without a guard. On Saturday morning went on a journey to the summit near the Rapids on the north side of the river, with the chief's brother as my guide, leaving the Canadian to take care of the tent and property. This took three days, and was one of the most laborious undertakings I ever experienced, the way was so rough, over dead wood, detached rocks, rivulets, &c. that very little paper could be carried. Indeed I was obliged to leave my blanket (which, on my route is all my bedding) at my first encampment about two-thirds up. My provision was 3 oz. tea, 1 lb. sugar, and four small biscuits. On the summit all the herbage is low shrub but chiefly herb plants. The second day I caught no fish, and at such a great altitude the only birds to be seen were hawks, eagles, vultures, &c. I was fortunate enough to kill one young white-headed eagle, which (then) I found very good eating. On the summit of the hill I slept one night. I made a small fire of grass and twigs and dried my clothes which were wet with perspiration and then laid myself down on the grass with my feet to the fire. I found it very cold and had to rise four times and walk to keep myself warm. Fortunately it was dry and a keen north wind prevented dew. On Monday evening at dusk I reached my tent at the village much fatigued and weak and found all things going on smoothly. Made a trip to the opposite side two days later, also to the summit of the hills, which I found of easier ascent, the only steep part near the top. My food during my stay was fresh salmon, without salt, pepper, or any other spice, with a very little biscuit and tea, which is a great luxury after a day's march.

Last night my Indian friend Cockqua arrived here from his tribe on the coast, and brought me three of the hats made on the English fashion, which I ordered when there in July; the fourth, which will have some initials wrought in it, is not finished, but will be sent by the other ship. I think them a good specimen of the ingenuity of the natives and particularly also being made by the little girl, twelve years old, spoken of when at the village. I paid one blanket (value 7s.) for them, the fourth included. We smoked; I gave him a dram and a few needles, beads, pins, and rings as a present for the little girl. Faithful to his proposition he brought me a large paper of seeds of *Vaccinium ovatum* in a perfect state, which I showed him when there, then in an unripe state. I have circulated notices among my Indian acquaintances to obtain it for me.

DAVID DOUGLAS, 1825 (pub. 1914)

Every gardener knows the fascination of the unknown, and when the ordinary plants are doing nicely there is a great temptation to be a little more venturesome. That is one of the excitements of gardening, but one which my husband did not share. He pretended not to see me with my nose in catalogues night after night, and though I always tried to intercept the postman when I was expecting plants, he always knew.

MARGERY FISH, 1956

Autumn brought the catalogues, of which, if my memory is true, there were at that time four only, emanating from Messrs Rivers, Paul, Lane, and Wood. Ah! had I studied my books at Oxford with half the zest with which I devoured these catalogues, what preeminence I might have won! I read, re-read, compared, and annotated those pages until my sisters asked sneeringly, 'What could I see in those stupid lists?' and prophesied an early softening of my brain. The youngest, I remember, to whom in an incautious moment I had exhibited my Masonic apron, 'felt sure that they came from that horrid lodge', and sniffed at them as though they smelt of sulphur. But to me, nevertheless, it was and has been from that day to this a never-failing amusement to study, as in a gallery, these portraits by different artists of Queen Rosa and her suite – a gratification like that which lovers feel as they gaze upon the likeness of their absent darling.

At last, and after as careful deliberation as though I had been some fond mamma who was engaged in choosing husbands for her daughters, with all the swells of Rotten Row to pick from, I made my 'purchaser's own selection', and sent my order to a neighbouring nurseryman, with quite as high an idea of its importance as though I were raising him to the peerage. My conviction was that no demand of similar magnitude (two dozen rose-trees!) had been previously made by any amateur, and that, when they were added to my existing stock of ten, they would be, as Mr Wombwell says of his menagerie, 'a magnificent and unrivalled collection'. I knew not then how the rose-lover's appetite grows with that it feeds on; I foresaw not the day when with 1,500 trees I should be sending my plate, like a distended schoolboy, for 'just a small slice more'.

S. REYNOLDS HOLE, 1909

Most of the People who sell the Trees and Plants in Stocks and other Markets are Fruiterers who understand no more of gardening than a Gardener does the making up of Compound Medicines of an Apothecary. They often tell us the plants will prosper, when there is no Reason or Hopes for their growing at all; for I and others have seen Plants which were to be sold in the Markets, that were as uncertain of growth as a Piece of Noah's Ark would be if we had it here to plant.

THOMAS FAIRCHILD, 1722

Henry James set this story [*The Aspern Papers*] in Venice. What follows is no more than is relevant to the garden. Two shy, mysterious American ladies, the Misses Bordereau, aunt and niece, live in a 'sequestered and dilapidated old palace' on a quiet canal. They possess a garden behind a high blank wall 'figured over with the patches that please a painter' and also some jealously-guarded secret papers which the hero intends to obtain. Seeing 'a few thin trees and the poles of certain rickety trellises' over the wall, he decides to use the garden as a pretext and to get himself taken in as a lodger.

He gains admission, meets the niece, 'a long pale person' of uncertain age, sees the garden from an upper window and decides that, though shabby, it has 'great capabilities'. He tells her he is doing literary work, *must* have a garden, must be in the open air and cannot live without flowers. It is already April, but he says: 'I'll put in a gardener. You shall have the sweetest flowers in Venice'; and that he would 'undertake that before another month was over the dear old house would be smothered with flowers'. The ancient aunt is rightly suspicious but needs money, so she offers him some empty rooms on the second floor at an extortionate rent. He takes them. Has he not already said that his tastes and habits are of the simplest, and added the irritating words: 'I live on flowers'?

Anyone with a knowledge of gardening may feel that Henry James has prepared a horticultural time-bomb here. His hero will never be able to supply enough flowers from a neglected garden to smother a large Venetian palace in the space of one month; indeed, he is already running two weeks behind schedule when 'Six weeks later, towards the middle of June' – so his first visit must have been at the end of April – he has furnished his rooms, moved in with a manservant and started to spend part of every day in the garden. The ladies will have nothing to do with him, he never sees them, but he had told himself that he would 'succeed by big nosegays. I would batter the old women with lilies – I would bombard their citadel with roses. Their door would have to yield to the pressure when a mound of fragrance should be heaped against it' – a tall order for any gardener. We sit back to watch him come unstuck.

Let us try to follow the timing. As soon as his rooms are arranged – and as this involves engaging his servant as well as the ordering and delivering of a boat-load of furniture, it must have taken at least a week – he 'surveyed the place with a clever expert and made terms' for having the garden put in order. But the 'Venetian capacity for dawdling is of the largest, and for a good many days unlimited litter was all my gardener had to show for his ministrations. There was a great digging of holes and carting about of earth.' After a while our schemer grows so impatient that he thinks of buying flowers – but he suspects the ladies will be watching through their shutters and contains himself. 'Finally, though the delay was long,' he 'perceived some appearances of bloom', he had 'had an arbour arranged' and, in it, worked and 'waited serenely enough till they multiplied'.

By July he was no nearer to obtaining the papers or even seeing the ladies, but as he sat in his arbour 'the bees droned in the flowers', though which he does not say, and he was spending the hot evenings floating in his gondola or at Florian's eating ices. Then, one evening, he comes back early, enters the garden's 'fragrant darkness' and finds the younger Miss Bordereau 'seated in one of the bowers'.

Now comes the bombshell, and it is not for the hero but for us. He says: 'I asked her why, since she thought my garden nice, she had never thanked me . . . for the flowers I had been sending up in such quantities for *the previous three weeks* . . . there had been . . . *a daily armful*' and he would have liked 'a word of recognition'.

Gardeners may well ask how this was possible. Even if work on the garden started immediately, although this seems unlikely among 'Venetians with their capacity for dawdling', if, say, the 'clever expert' was able to find an unemployed gardener by the middle of May, the process of taming a wilderness can take weeks, and we have been told that the garden was 'a tangled enclosure' and had been 'brutally neglected'. We are not told how big it was nor what plants had survived from earlier days, although Miss Bordereau had said 'we've a few but they're very common', while the reference to rickety trellises leads us to suppose that there were at least some climbing roses and jasmine. There was only one gardener to clear the 'unlimited litter', dig his holes, 'cart earth about' and get his plants established – and he may also have had to 'arrange' the arbour. How on earth had he done it?

With his orders to provide 'big nosegays' we can imagine him, by the end of May, working from dawn to dusk, feverishly preparing the soil and sowing annuals – work that should have been done in March or April – and planting anything he could find that would produce flowers suitable for cutting. What would he choose? 'Digging holes' suggests that he put in a number of shrubs, but which? So many shrubs that are useful for flower arranging have finished blooming by July: lilacs, philadelphus and so on, likewise wisteria and blossoming trees – and few shrubs produce much in their first year anyway. Would oleanders or hibiscus have fitted the bill? The irises which grow so well in Italy would also be over, with the peonies, oriental poppies, lupins and the rest of the first flush of perennial flowers. He must have relied heavily on whatever was supplied in pots by the local shops and stalls or perhaps by nurserymen in the Veneto. But we are speaking of late Victorian Venice, not England, today. Did they sell bedding plants in boxes? Could he, at this late stage, obtain sweet peas, carnations, stocks, lilies, potted or boxed and ready to flower? In England sweet peas should be sown, and gladiolus corms planted, in March. Would these be available in Venice as late as May? Sweet Williams should have been planted the previous autumn, and are hardly Italian flowers.

Classic Italian gardens are not known for masses of flower colour but for architectural design, topiary, stonework, ironwork and water, the contrast of light and shade, for cypresses, yew, bay, box and marble, and while Venice has many pots and window boxes these are mainly of small plants, trailing geraniums, marigolds and the like. Such colour as there is in Italian gardens seems, at least in these days, to come, apart from roses, mainly from such things as azaleas, fuchsias, geraniums, Paris daisies and petunias grown in containers that are set out at strategic points as they come into bloom. You could not pick armfuls of flowers from them, and if you did they would take weeks to recover, while a gladiolus or lily, having bloomed, has shot its bolt and can be picked no more.

More is to come. After a talk with the younger lady during which they 'wandered two or three times round the garden' he dares to mention the all-important papers. She takes fright, hurries away and does not reappear, and after four or five days he tells the gardener 'to stop the "floral tributes"'. The cutting of armfuls of flowers has lasted for very nearly *four weeks*. In a later scene, between our hero and the ancient aunt, she thanks him for the flowers: 'You sent us so many . . . I suppose you know you can sell them – those you don't use.' And he replies: 'My gardener disposes of them and I ask no questions.' It is quite a jolt to realize that Henry James saw flowers as something to be used or disposed of, and that he imagined that once a supply of flowers – the only ones he names are lilies and roses – has been turned on, it flows like a fountain that cannot be turned off. His hero even turns to the younger Miss Bordereau and invites her to 'come into the garden and pick them; come as often as you like; come every day. The flowers are all for you.'

Even if the garden were fairly large – and he says at one point that he 'wandered about the alleys smoking cigar after cigar' – it must have been remarkable. This is a famous novel, written by a master, and the plot takes many fascinating turns: but we are amazed to find the niece still picking flowers weeks later, her hands full of 'admirable roses'. Surely most of the roses grown at that time bloomed in June and early July only. What repeat-flowering roses would we have found in Venice then?

Could it have been done? Or, and I may well be wrong, are there only two possible explanations: either that gardener, whose name is never mentioned, and who was never given a word of praise, was an unsung hero of horticultural literature, a genius who given an acre in England could have supplied Covent Garden; or – dare I suggest it? – fear it may be the case – Henry James, that legend, that great word-smith and story-teller, was as horrid as his hero who, in all the long days that the novel records, never touches a flower, never lends a hand by so much as dead-heading one rose, and, like his hero who pretended to love flowers so much but knew only two by name, was a townee and a beast, and unworthy to write about flowers?

NANCY-MARY GOODALL, 1987

Know, stranger, ere thou pass, beneath this stone
Lye John Tradescant, grandsire, father, son,
The last dy'd in his spring, the other two
Liv'd till they had travell'd Orb and Nature through,
As by their choice Collections may appear,
Of what is rare, in land, in sea, in air,
Whilst they (as Homer's Iliad in a nut)
A world of wonders in one closet shut,
These famous Antiquarians that had been
Both Gardiners to the Rose and Lily Queen,
Transplanted now themselves, sleep here & when
Angels shall with their trumpets waken men,
And fire shall purge the world, these three shall rise
And change this Garden then for Paradise.

Epitaph on the tomb of John Tradescant the Elder
(c.1570–1638), his son John (1608–1662)
and grandson John III (1633–1652), in the
churchyard of St Mary-at-Lambeth, London

CHAPTER THREE

Work in the Garden

I left London by the Comet Coach for Chesterfield; and arrived at Chatsworth at half-past four o'clock in the morning of the ninth of May, 1826. As no person was to be seen at that early hour, I got over the greenhouse gate by the old covered way, explored the pleasure grounds and looked round the outside of the house. I then went down to the kitchen gardens, scaled the outside wall and saw the whole of the place, set the men to work there at six o'clock; then returned to Chatsworth and got Thomas Weldon to play me the waterworks and afterwards went to breakfast with poor dear Mrs Gregory and her niece; the latter fell in love with me and I with her, and thus completed my first morning's work at Chatsworth before nine o'clock.

SIR JOSEPH PAXTON, 1826

There is a point, usually toward the middle of March, when a stroll around the garden – just to escape the house, just to get some air – takes on a quite different quality from the ambles of deep winter. The ground is still frozen hard, the tender perennials and heathers are still swathed in evergreen boughs, and the boxwoods are still sleeping under their ungainly crates. There is detritus everywhere: sodden perennials and annuals that the autumn left no time for cutting away, rotting leaves swirled among the shrubs, snow compacted into stubborn, heel-bruising ridges of ice, and even less mentionable things – a winter's leavings from household pets, perhaps even a full vacuum cleaner bag put out the back door and forgotten beneath an obscuring fall of snow. There is little that is beautiful, and such as there is – the somber forms of evergreens, the courage of witch hazels, the enameled stems of willows and shrubby dogwoods – has been celebrated too much, fed on for too long to be of continuing interest now. But still, in the middle of this most dreadful of gardening months, one feels a difference in the garden, and in one's own personal chemistry. It is not so much in the behavior of the plants, though to very attentive eyes the willow buds are certainly swelling, and the snow around plantings of galanthus begins to be pierced with curious spikes of celadon. But the tap one feels on one's shoulder, the rise in one's heart, is from a sudden gust of warm, moist air. The direction of things has changed. Tentatively, shyly, like the first promise of very young love, the new year has begun.

JOE ECK AND WAYNE WINTERROWD, 1995

At this time [1786] received a letter from the Duchess of Rhohan Chabot who has a country house at St Mandé to go to speak with her relative to the Making a garden; went and met the Duchess at St Mandé, the place a narrow strip of garden with some lime trees to the left planted in little Salles and narrow walks and the right side a sorte of Kitchen garden runing allong the wall; after examining the ground I asked the Duchess to make her nearly a small project of what I meant to do which with a Pencil was soon done and seemed to please her very much; she said that she had applied to Gabriel Thouin but he could give her no ideas of beauty or simplicity; however she wanted me to conduct the work and I asked one of her domisticks rather than the Gardner, so she adopted one Cleremont a very intiligent Man. The

Duchess is a good friendly woman, she said she wished it done before the Duke should know anything of the changes as she wished to give him an agreeable surprize so that the work was begun and ended very expeditiously in about three Months; so that she invited the Duke one day to dine and when he arrived he could hardly be made to beleive it was possible to have changed his place to what it then was; he loaded me with compliments and found every thing so fine; at this time the Duchess aplied to Me for a gardener but as she said 'Mr B. you know how my house is composed, I would desire something but perhaps you may laugh.' 'I ask you pardon' Said I 'dont think I would do such a thing.' She told me that she desired a good Gardener and a good Catholick at the same time. I told her I should examin. However I had at the same time a young man a German who was a Protestant whome I recomended and lived with them and pleased them well untill they left him at the Revolution; however the gardens in this country let them be ever so well done the keeping and fine grass in England is what is frequently more to be admired than the designe which is frequently little observed, allthough they might varie their gardens more than they do and frequently their Slopes badly executed. Every place if rightly understood may be rendered agreeable in observing the ground where nature has frequently made Beautys which is not observed by those that has the direction. About this time the Duchess of Rhohan desired me to go with her to see a garden belonging to one of her relations who was the Spanish Ambassador; this was a garden at his hottel rue de L'University the place tolerable large and planted in the Ancient manner with Straight walks of Limes; after several things proposed the Ambassador found pretty but what he said was so absurd that he wished to change the whole without changing any thing; such reasoning from the Ambassador made one laugh and the Duchess calling me aside said 'O le vilain homme!' so I told him not to think any more of the Changes as I found his garden well and I agreed not to change any thing, only that I wished him good day and so we parted and I returned with the Duchess to her hotel who exclamed much against the absurdity of her relation; however as I said we must overlook those absurditys allthough they are rather provoking and loss of time which those people thinks little of. About this time the Queen brought over an English gardener who was to make the grass at Trianon as fine as in England but he soon was lost and returned in disgrace to England, as he was a Man of no Genius and as far as I could see knew very little, as he had a formidable rival in Richard who soon undermined him and as those gardens is allways under the direction of the Architect and as this man was not placed by Mr Micque, who was Architect and pretended to make and conduct the Gardens which was done with more expence than taste.

THOMAS BLAIKIE, 1786

My father mistrusted gardeners – they dig up all one's pet plants, he avowed – and would not have one anywhere about the place, so always I was commandeered to do the weeding and clearing that bored him. 'When I grow up I'll never, never, never have a garden,' I resolved, as day after day I uprooted daisies from the tennis court or tidied the edges of the paths. And I meant it. But now that there is no force to command me but the needs of the garden itself, I am happy with it.

CLARE LEIGHTON, 1935

If you want to get on in the upper reaches of the gardening world – well, perhaps not quite upper; just . . . not quite – you simply must not try to question the faith. An example of this is the story of the painting of tree wounds. For some time it was thought that the stumps of severed branches should have their surfaces painted with a fungicidal paint. The arguments for it were convincing enough for professionals to recommend and carry out the practice. Later the same professionals, who, after all, do the job every day, noted that as time passed the painted stumps decayed more rapidly and dangerously than those left unpainted. By now, however, the guardians of the arboricultural mysteries had taken stump painting to their collective bosom. It had become as indispensable to the perceived health of trees as purging had once been to the treatment of malaria. To this day, it is demanded of tree surgeons that they paint the stumps of branches they have removed; and they do so, otherwise they would lose business. However, they now use cheap household emulsion from the Penny Bazaar, in the hope that it will wash off only after their bill has been paid.

JOHN KELLY, 1992

The massacre of dandelions is a peculiarly satisfying occupation, a harmless and comforting outlet for the destructive element in our natures. It should be available as a safety valve for everybody. Last May, when the dandelions were at their height, we were visited by a friend whose father had just died; she was discordant and hurt, and life to her was unrhythmic. With visible release she dashed into the orchard to slash at the dandelions; as she destroyed them her discords were resolved. After two days of weed slaughtering her face was calm. The garden had healed her.

CLARE LEIGHTON, 1935

When my grandfather was ninety-nine he decided to dig an asparagus trench in his garden. The chauffeur, a man named Pepper, was summoned from one of his endless card games with the cook, and told to mark out the site of the ambitious project, which ran across part of the lawn and bisected a flower bed:

'He will overtax his strength,' said my grandmother, as the two men emerged from the tool-shed carrying spade and pick. She had been saying this, almost daily, for many years – though for fewer than one might imagine, for she had been married from the schoolroom, and was only sixty-seven. I was twelve.

My grandfather attacked his trench for an hour every day before tea. Very soon it became apparent that his ideas about asparagus were original. The trench advanced very little, but every day grew deeper. After about a week, hacking at a stone, the old man sprained his wrist. Operations were suspended.

'He doesn't like asparagus,' said my grandmother. 'He was trying to discover what the trenches were like for your father during the war.'

She was right, of course. Pepper was instructed to cut up a stepladder and put it against the wall of the trench. From then on my grandfather spent a few minutes every afternoon on the firing step, and once, when I was playing not far away, I received a clod of earth on the head. A grenade, of course.

JOHN LODWICK, 1960

January:

In this moneth Graff [graft] in the cleft, decrease of the Moone; & towards the end thereof prune wall fruit, 'til the sap rises briskly, especialy finish cutting your Vines.

February:

Naile yet & prune: sow all sorts of Kernels, towards ye later end Melons & rare seedes on the Hot-bed.

March:

Sow Endive, Succory, Chervil, Sellerie, purselan (which you may also continue sowing all the summer to have tender) leeks, Beetes, parsneps, salsifix, skirrits, Turneps, &c. and now Cherish and Earth-up your flowers, and set stakes to the tallest: sow also lettuce.

Aprill:

Set Artichock-slips, transplant cabages, sow Lettuce, clip hedges, & greenes; & sow the seedes of all hot sweete-herbs & plants.

May:

Bring forth of the Greene-house the Oranges, Lemons and most tender Ever-greenes, trim and refresh them, placing them in shade a fortnight, by degrees accostuming them to the sunn: sow also cabbage-seedes, Lettuce, French-beanes, Harricos &c.

June:

Sow Lettuce, Raddish –

July:

Sow Lettuce, remove Cabbage-plants, Lay ever-greens, and transplant such as are rooted, do this about St Jamestide.

August:

Sow Cabbages, Carrots, Turneps, purselan – Innoculate oranges and other rare plants: Begin to prune over shady shootes of the Spring, yet so as not to expose the Fruit.

September:

Sow Lettuce, Spinach – plant primroses, violets & such fibrous rootes.

October:

Sow Lettuce, Alaternus, phillyrea seedes, Kirnels &c. and now begin to secure & by little & little, as the season proves, withdraw your choicer & tender Greenes & prepare them for the Greene house.

November:

Trench & prepare ground with compost – sow as yet all sorts of greenes.

December:

Carry, & spread dung & compost.

JOHN EVELYN, 1686 (pub. 1932)

The Gardiner should walke aboute the whole Gardens every Monday-morning duely, not omitting the least corner, and so observe what Flowers or Trees & plants want staking, binding and redressing, watering, or are in danger; especialy after greate stormes, & high winds and then immediately to reforme, establish, shade, water &c what he finds amisse, before he go about any other work.

Monday: . . . Early, before the deaw be off in Mowing season, and as his grasse is growne too high (that is, if any daiysie or like appeare) he is to cut the grasse of the greate Court, & roll all the gravell: having rolled also the carpet, the *Saturday* night before, and this Monday evening the upper Terrace & lower.

Tuesday: Mow the upper & Lower Terrace grasse walkes: rolling the grasse of the Fountaine & Greene-house garden grasse with the grasse walkes of both Groves that evening: and Gravell walkes.

Wednesday: Mow the Fountaine & Greene-house Garden, with both the Groves, & Roll in the evening the long middle grasse walk to the Iland.

Thursday: Mow the Middle grasse walk to the Iland: Roll the Broad Gravell-walke, The Holy-hedge, and grove gravell-walkes: & this evening roll the three crosse walkes from the Ort-yarde to the Iland moate.

Friday: Mow the three former crosse grasse walkes.

Saturday: Evening, Roll the Bowling Greene: & Court.

. . . And thus alternatively may all the Grasse & Walkes be rolled, & cut once a fortnight, with ease: that is the grasse every 15 dayes, & the gravell rolled twice every six dayes.

Weeder: Note that whilst the Gardener rolls or Mowes, the Weder is to sweepe & clense in the same method, and never to be taken from that work 'til she have finished: first the gravell walkes & flower-bordures; then the kitchin-gardens; to go over all this she is allowed One moneth every three-moneths, with the Gardiners assistance of the haw, & rough digging; where curious hand-weeding is lesse necessary.

Every fortnight looke on Saturday to your seede and roote boxes, to aire & preserve them from mouldinesse & vermine.

Looke every moneth (the last day of it) & see in what state the Bee-hives are: and every day, about noone if the weather be warme, and the Bees hang out for swarmes; having y^r hives prepar'd & ready dressed.

The Tooles are to be carried into the Toole-house, and all other instruments set in their places, every night when you leave work: & in wett weather you are to clense, sharpen, & repaire them.

The heapes of Dung, & Magazines of Mould &c: are to be stirred once every quarter, the first weeke.

In Aprill, Mid-August, clip Cypresse, Box, & generally most ever-greene hedges: & closes, as quick-setts.

Prune standard-fruit & Mural Trees the later end of July, & begin-ning of August for the second spring: Vines in January & exuberant branches that hinder the fruite ripning in June.

The Gardner, is every night to aske what Rootes, sallading, garnish-ing, &c will be used the next day, which he is accordingly to bring to the Cook in the morning; and therefore from time to time to informe her what garden provision & fruite is ripe and in season to be spent.

He is also to Gather, & bring in to the House-Keeper all such Fruit of Apples, peares, quinces, Cherrys, Grapes, peaches, Abricots, Mul-beries, strawberry, Rasberies, Corinths, Cornelians, Nutts, Plums, & generally all sort of Fruite, as the seasone ripens them, gathering all the windfalls by themselves: That they may be immediately spent, or reserved in the Fruite & store-house.

He may not dispose of any the above said Fruite nor sell any Artichock, Cabbages, Aspargus, Melons, strawberries, Rasberies, Wall, or standard & dwarfe fruite, Roses, Violets, Cloves, or any Greenes, or other flowers or plants, without first asking, and having leave of his Master or Mistress; nor till there be sufficient of all garden furniture for the Grounds stock and families use.

He is to give his Mistris notice when any Fruites, Rootes, Flowers, or plants under his care are fit to be spent, reserved, cutt, dried, & to be gathered for the still house and like uses, & to receive her directions.

He is, when any Tooles are broaken or worn out, to bring the

Instrument so unserviceable to his Master and shew it before another be bought.

Let him for all these observations, continualy reade and consult my Gardiners Almanac & Discourse of Earth.

JOHN EVELYN, 1686 (pub. 1932)

I left my own garden yesterday, and went over to where Polly was getting the weeds out of one of her flower-beds. She was working away at the bed with a little hoe. Whether women ought to have the ballot or not (and I have a decided opinion on that point, which I should here plainly give, did I not fear that it would injure my agricultural influence), I am compelled to say that this was rather helpless hoeing. It was patient, conscientious, even pathetic hoeing; but it was neither effective nor finished. When completed, the bed looked somewhat as if a hen had scratched it: there was that touching unevenness about it. I think no one could look at it and not be affected. To be sure, Polly smoothed it off with a rake, and asked me if it wasn't nice; and I said it was. It was not a favourable time for me to explain the difference between puttering hoeing, and the broad, free sweep of the instrument, which kills the weeds, spares the plants, and loosens the soil without leaving it in holes and hills. But, after all, as life is constituted, I think more of Polly's honest and anxious care of her plants than of the most finished gardening in the world.

CHARLES DUDLEY WARNER, 1876

When mother in her young days was telling her fortune from cards she always whispered over one pile: 'What am I treading on?' Then I could not understand why she was so interested in what she was treading on. Only after very many years did it begin to dawn on me. I discovered that I was treading on the earth.

In fact, one does not care what one is treading on; one rushes somewhere like mad, and at most one notices what beautiful clouds there are, or what a beautiful horizon it is, or how beautifully blue the hills are; but one does not look under one's feet to note and praise the beautiful soil that is there. You must have a garden, though it be no bigger than a pocket-handkerchief; you must have one bed at least to know what you are treading on. Then, dear friend, you will see that not even clouds are so diverse, so beautiful, and terrible as the soil under your feet. You will know the soil as sour, tough, clayey, cold, stony, and rotten; you will recognize the mould puffy like pastry, warm, light, and good like bread, and you will say of this that it is beautiful, just as you say so of women or of clouds. You will feel a strange and sensual pleasure if your stick runs a yard deep into the puffy and crumbling soil, or if you crush a clod in your fingers to taste its airy and tepid warmth.

And if you have no appreciation for this strange beauty, let fate bestow upon you a couple of rods of clay – clay like lead, squelching and primeval clay out of which coldness oozes; which yields under the spade like chewing-gum, which bakes in the sun and gets sour in the shade; ill-tempered, unmalleable, greasy, and sticky like plaster of Paris, slippery like a snake, and dry like a brick, impermeable like tin, and heavy like lead. And now smash it with a pick-axe, cut it with a spade, break it with a hammer, turn it over and labour, cursing aloud and lamenting.

Then, you will understand the animosity and callousness of dead and sterile matter which ever did defend itself, and still does, against becoming a soil of life; and you will realize what a terrible fight life must have undergone, inch by inch, to take root in the soil of the earth, whether that life be called vegetation or man.

And then you will know that you must give more to the soil than you take away; you must make it friable and fertile with lime, and temper it with warm manure, lighten it with ashes, and saturate it with air and sunshine. Then the baked clay disintegrates and crumbles as if it breathed in silence; it breaks down under the spade with surprising readiness; it is warm and malleable in the hand; it is tamed. I tell you, to tame a couple of rods of soil is a great victory. Now it lies there, workable, crumbly, and humid; you would like to take it and rub it all between your thumb and finger, to assure yourself of your victory; you think no more of what you will sow in it. Is it not beautiful enough, this dark and airy soil? Is it not more beautiful than a bed of pansies or carrots? You are almost jealous of the vegetation which will take hold of this noble and humane work which is called the soil.

And from that time on you will not go over the earth unconscious of what you are treading on. You will try with your hand and stick every heap of clay, and every patch in a field, just as some other men look at stars, at people, or violets; you will burst into enthusiasm over the black humus, fondly rub the smooth woodland leafmould, balance in your hand the compact sod and weigh the feathery peat. O Lord! you will say, I should like to have a wagon of this; and heavens! a cartload of this leafmould would do me good; and this humus here for putting on the top, and here a couple of those cow pancakes, and a little bit of that river sand, and some rings of these rotten wood stumps, and here a bit of sludge from the stream, and sweepings from the road would not be bad either, would they? and still some phosphate and horn shavings, but this beautiful arable soil would also suit me. Great Scott! There are soils as fat as bacon, light as feathers, crumbly like a cake, blond or black, dry or inflated with damp; all these are diverse and noble kinds of beauty; while all that is greasy, cloddy, wet, tough, cold, and sterile is ugly and rotten, unredeemed matter, given to man for a curse; and it is as ugly as the coldness, callousness, and malice of human souls.

KAREL ČAPEK, 1931

Whoever begins a garden, ought in the first place, and above all, to consider the soil ... Perhaps Varro's judgment upon this case is the wisest and the best; for to one that asked him, what he should do if his father or ancestors had left him a seat in an ill air, or upon an ill soil? He answered, Why sell it, and buy another in good. But what if I cannot get half the worth? Why then take a quarter; but however sell it for anything, rather than live upon it.

SIR WILLIAM TEMPLE, 1685 (pub. 1692)

Everyone has a theory about composting. I got my recipe from an American book, and I find it works out well. All green stuff is put in a heap to rot down. Perennial weeds and evergreen material are not used, but everything else, including kitchen refuse, grass cuttings, great mountains of nepeta, aubrieta, Michaelmas daisies and all the other herbaceous things that are cut down. For the kitchen refuse I keep a big brown pot (commonly known as 'the gash') on the window sill behind the sink, and into it go all the tea-leaves, apple peels, onion skins and coffee grounds. Also crushed egg shells. Walter made a great fuss about the egg shells, he disliked them so and contended that it was silly to bother about them when I could get all the lime I wanted for a few pence. But I think my plants enjoy a mixed diet and I would not deny them little tit-bits of shell, but I did see that they were crushed very finely so that they did not intrude too forcibly on my lord's eye. I noticed great mounds of coffee grounds at Kew, so I know I am on firm ground there, and as for tea-leaves, you have only to see what emptying the teapot does to a wilting plant. I have known trees and shrubs brought back from the dead by having tea and tea-leaves administered to them after every meal, and I am sure one reason why Madonna lilies thrive in cottage gardens and not in ours is because they get tea and washing-up water and all manner of good things given to them. I don't like the messiness of tea-leaves thrown on the flower beds, but I use them and the tea that is left in the teapot in the compost. The liquid is particularly good, in fact a necessity, for in very dry weather the compost heap needs generous watering to speed decay.

I leave this heap until it is quite brown, and then I combine it with other ingredients to stand again. How long one leaves it depends on the speed of decomposition and the supply of material. I have now built up such reserves that the making of the final heap is done in the winter, and the following autumn I have a plentiful supply of super nourishment with which to enrich the garden.

My final heap is made in four layers, repeated until all the material is used up. First there is a generous layer of my rotted compost, then an equal depth of farmyard manure. This is then covered with earth and thickly dusted with wood ash. Pipes are inserted vertically at regular intervals down the heap as it is being built, so that it shall be ventilated. I like to use very young manure so that a high temperature kills any weed seeds that may be lurking in the compost. As my natural soil is clay, and such heavy clay that it doesn't change a bit during its year's sojourn in the heap but comes out a soggy solid mass, like a layer of marzipan in an Easter cake, I am now using sand instead of soil in the heap, but finish with grass tufts skimmed from the vegetable garden. These turned upside down seal the heap and keep in the heat to do its work thoroughly.

MARGERY FISH, 1956

Culture also [is] of very great Importance: And this is often discover'd in the taste and consequently in the Goodness of such Plants and *Salleting* [salads], as are Rais'd and brought us fresh out of the Country, compar'd with those which the Avarice of the *Gardiner*, or Luxury rather of the Age, tempts them to force and *Resuscitate* of the most desirable and delicious Plants.

It is certain . . . that about populous Cities, where Grounds are over-forc'd for Fruit and early *Salleting*, nothing is more unwholsom: Men in the Country look so much more healthy and fresh; and commonly are longer liv'd than those who dwell in the Middle and Skirts of vast and crowded Cities, inviron'd with rotten Dung, loathsome and common Lay Stalls; whose noisome Steams, wafted by the Wind, poison and infect the ambient Air and vital Spirits, with those pernicious Exhalations, and Materials of which they make the *Hot Beds* for the raising those *Præcoces* indeed, and forward Plants and Roots for the wanton Palate; but which being corrupt in the Original, cannot but produce malignant and ill Effects to those who feed upon them . . . There being nothing so proper for *Sallet Herbs* and other *Edule Plants*, as the Genial and Natural Mould, impregnate, and enrich'd with well-digested Compost (when requisite) without any Mixture of Garbage, odious Carrion, and other filthy Ordure, not half consum'd and ventilated and indeed reduc'd to the next Disposition of Earth it self, as it should be; and that in Sweet, Rising, Aery and moderately Perflatile Grounds; where not only *Plants* but *Men* do last, and live much longer. Nor doubt I, but that every body would prefer Corn, and other Grain rais'd from *Marle, Chalk, Lime*, and other sweet Soil and Amendments, before that which is produc'd from the *Dunghil* only.

JOHN EVELYN, 1699

It was some six months after I had arranged to buy the Villa Taranto before I could return to Italy and start the work of reconstruction that I had been busy planning in my mind. In June 1931 I arrived in Pallanza and began on the first steps of my long-range campaign. Both house and garden had to be remodelled, and my first task was to find the right workmen for my ambitious schemes.

Only one gardener had been left at the villa by the former owner, but I had the good fortune to find another who knew something about gardening. For the first year I was my own head gardener, and had only six Italians who knew something (really very little) about gardening. The rest were labourers. And what a host of them there were! One set of men was rebuilding the villa and another larger group was at work on the gardens. To see the hundreds of men streaming out of the gates at the end of the day was really like watching a factory after the blowing of the stop-work siren.

We attacked the woods first, and we cut and cut and cut, getting rid of masses of poorly grown and deformed chestnuts and some un-attractive pines, together with quantities of bamboos. I think I kept only a score or so of trees and shrubs altogether. These included beeches – weeping, copper and cut leaf – some very fine old castaneas, a few conifers and several *Magnolia grandiflora*, a few very old camellias and several enormous specimens of *Chimonanthus fragrans*.

Before long I had installed a Deccaville railway, the type we had used in Salonika in the 1914–18 War for bringing up supplies to the troops. This was invaluable for carting stones and soil and for remov-ing old trees. Some of the trees were cut and stacked for firewood for the villa, and thousands of tons were distributed to people in the district.

When the trees had been removed, I began planting silver birches which I had bought in Italy. These birches were about twelve feet high, and they established themselves promptly in the rich soil and grew extraordinarily quickly. Thus the first stages of my plan took form.

Another urgent job was to replace the Italian gardens on the east side of the house. A big terrazza was constructed on this side of the villa, as I thought that a lawn, surrounded with flower-beds and trees framing a glimpse of the lake, would be a pleasant feature to look out upon from the house.

Adequate drainage was a task which could not be postponed, and roads and retaining walls had to be built. These walls were particularly important as with so much clearing going on there was constant risk of land-slides on some of the steeply-sloping parts of the gardens. The whole place seethed with industry as so many urgent works were being carried out at one and the same time.

I soon decided that I could no longer get on without a head gardener. I wanted an Englishman who had been at the Royal Botanic Gardens, Kew – one who could speak some Italian. Remembering La Mortola, Sir Cecil Hanbury's garden at Ventimiglia, where they had Kew students, I wrote to Sir Cecil, who very kindly recommended me one. We worked together for some time, but he then returned to England to get married.

I made enquiries for several months, but could not find a suitable person. Suddenly an old friend, Captain Jack Tremayne of Heligan in Cornwall, who owned the Villa Boccanegra at Ventimiglia, wrote to me to say that he had found the very man I wanted; he asked me to come and see him. There I met Mr Henry Cocker, and arranged with him to come to Pallanza to see the Villa Taranto. Mr Cocker had been one of the most promising Kew students as well as an exchange student at La Mortola. I engaged him in 1934. He has been with me ever since and is now Superintendent of the gardens. Henry Cocker has become a great friend of mine. I value his loyalty and his help. His knowledge of horticulture is extremely wide, and I was very pleased when in 1950 he was made an Associate of Honour of The Royal Horticultural Society. He has done very valuable work in Italy, and acts as a judge at horticultural shows in Rome, San Remo and Turin.

NEIL MCEACHARN, 1954

A well-kept garden makes credible to modern eyes the antique fable of an unspoiled world – a world where gaiety knows no eclipse, and winter and rough weather are held at bay. In this secluded spot the seasons slip by unawares. The year's passing-bell is ignored. Decay is cheated of its prize. The invading loss of cold, or wind, or rain – the litter of battered Nature – the 'petals from blown roses on the grass' – the pathos of dead boughs and mouldering leaves, the blighted bloom and broken promise of the spring, autumn's rust or winter's wreckage are, if gardeners be brisk sons of Adam, instantly huddled out of sight, so that, come when you may, the place wears a mask of steady brightness; each month has its new dress, its fresh counterfeit of permanence, its new display of flowers or foliage, as pleasing, if not so lustrous as the last, that serves in turn to prolong the illusion and to conceal the secret irony and fond assumption of the thing.

JOHN D. SEDDING, 1890

D eadheading is a most important part of gardening. It isn't only from the point of tidiness that one should remove spent flowers. A plant will go on flowering over much longer periods if every dead bloom is removed at once. Kept in a state of frustrated motherhood it will go on producing flowers in the hope of being allowed to set seed and thus reproduce itself. I often get three flowerings on Canterbury Bells by persistent deadheading, and I even deadhead my naturalized daffodils so that they do not deteriorate. I have some old swords and I keep one sharpened for this job. One can slash off a lot of heads in a very short time.

MARGERY FISH, 1956

30 Oct., 1682. Being my birthday, and I now entering my greate climacterical of 63, after serious recollections of the yeares past, giving Almighty God thanks for all his mercifull preservations and forbearance, begging pardon for my sinns and unworthinesse, and his blessing and mercy on me the yeare entering, I went with my Lady Fox to survey her building, and give some directions for the garden at Chiswick; the architect is Mr May; somewhat heavy and thick and not so well understood; the garden much too narrow, the place without water, neere an highway, and neere another great house of my Lord Burlington, little land about it, so that I wonder at the expence; but women will have their will.

JOHN EVELYN, 1682 (pub. 1818–19)

After you have finished your borders, knots or quarters as occasion serves, then you are to make your Walks, first pitching a range of Stakes just in the middle, only about a pole or more asunder . . . and according to the nature of the level; but in case your Walks are very wide, then you are to have three ranges of stakes for the more exact laying of your walks, but so not make the fall on each side of your walk too great or sharp, but rather a fine, almost undiscernable fall, except your ground be very apt to lye wet, so great a fall is both unhandsom and uneasie for such as wear high heel'd shooes; after you have staked your Walk, or before, you are to dig and level your walk with its own earth or gravel; but in case the earth lie too high or be too good, it will be your best way to carry some of it away for a better use, than only to walk on, and in the room thereof to bring or lay either Gravel or Chalk, or the like, and when you have rough levelled your walk, let it be well trodden or beat, that it may not settle unequally, after you have finished; and then having rough levelled as aforesaid, then lay on your fine Gravel if you have two sorts, and remember that if you lay your fine Gravel of a good considerable thickness, you may once a year or once in two years new break over your Walks, and flourish them over with a little or no charge, whereby your Walks is first to spread and lay your Gravel as it ought to lie, and then to tread it all over alike, and then to rake it again, and then with a beater made of a piece of Plank to settle it all over, not bringing your rowl upon your Walk until it be well settled, except it be a light wooden rowl, least you cause your Walks to lie in whamps; and after you have rowled your Walks once or twice in a place, at length it will be a good way to rowl them over-thwart with your light rowl, your Walks will shape the better; also in case the weather be very dry, it will be a good way after you have shaped your Walks, to water them well with a watering pot; and then when they are so dried as that you may walk on them, you may beat or rowl them as you see cause.

Take notice, that if your gravel be of a very lean nature; and have no earthy substance to cause it to bind, you may amend that fault by mixing a quantity of short lome or clay equally amongst your Gravel, but be careful, you do not put too much, lest you make a worse fault, therefore use the mean; much more might be said as to the making of Walks, as to the prevention of worms-casting, for which some use store of Salt or Soot, and also for prevention of moss, *&c.* but being strained for want of time, let this suffice.

LEONARD MEAGER, 1688

It will be requisite to have in the middle of one side of [the] Flower-garden a handsome Octangular Somer-house, roofed every way, and finely painted with Landskips, and other conceits, furnished with Seats about, and a Table in the middle; which serveth not onely for delight and entertainment, to fit in, and behold the beauties of the Flowers, but for many other necessary purposes; as to put the Roots of Tulips and other Flowers in, as they are taken up, upon Papers, with the names upon them, until they be dried, that they may be wrapped up and put in Boxes; for writing the names, both in planting and taking up, of all Flowers, in order as you dispose them; for shelter, in case of a sudden shower of rain; and divers other purposes you will find this House to be fit for, which is indeed a thing so necessary, that it cannot (with any convenience) be wanting, and therefore ought to be considered in the setting up of the Walls, wherein it is to be placed, so that it come not further into the Garden, than the just breadth of the Border, for putting of it out of square.

JOHN REA, 1676

The seats in a garden or pleasure-ground are generally purchased ready made; but an agreeable variety may be occasionally produced, by having the stump of an old tree formed into a seat, and twining ivy, and creeping flowering shrubs, round it. Where it is an object to save trouble, a plant of the Virginian creeper may be planted with one of the giant ivy; and if both are left to nature, the effect will be very good, as the brilliant deep red of the Virginian creeper in autumn will be relieved by the dark green of the ivy. A few moveable seats – one to wheel about from one part of the garden to another, and another of the new folding kind, imported from Norway by Charlwood, and sold at 3s. 6d. each, are very convenient. Where there is a terrace, a seat may be erected at each end, of wood, but of a somewhat massive design, and painted white, being strewed while the paint is wet with very fine sand, which will make it a good imitation of stone. Seats may also be decorated by nailing on a wooden frame-work of hazel rods with the bark on, which have been stained of different colours, and then varnished. These rods are arranged in a pattern, and I have seen the effect of a landscape produced; but it appears a kind of decoration that is in very doubtful taste, or at least one that it requires great judgment to manage properly.

JANE LOUDON, 1841

The most commendable inclosure for every Garden plot, is a quick-set hedge, made with brambles and white thorne: but the stronger and more defensive hedge is the same, which the singular *Democritus* in his Greek instructions of Husbandrie (that wrote long before *Columella*, and *Palladius Rutilius*) cunningly uttereth, and the same with ease and smal cost after this manner: Gather saith he, in a due season of the yeare, the seeds found in the red berries of the biggest and highest Briers (which by a more common name with us, are called the wild Eglantine Briers) the thorow ripe seeds of the brambles (running low by the ground) the ripe seeds of the white Thorne, and to these both the ripe Berries of the Goose-berry and Barberry trees; this done, mix and steepe for a time, all the Berries and seeds in the bending meale of Tares, unto the thickness of Honey: the same mixture lay diligently into old and untwisted Ship or Wel-ropes, or other long worne ropes, and fittered or broken into short pieces, being in a manner starke rotten, in such order, that the seeds bestowed or couched within the soft haires of them, may be preserved and defended from the cold, unto the beginning of the spring. At which time where you be minded that the inclosure or hedge shall runne and spring up, there digge in handsome manner, two smal furrows, and these either two or three foot asunder, and a mans foot and a halfe deep: into which lay your ropes with the seeds ...

THOMAS HILL, 1577

'Pleaching' is probably the best abused of the many iniquities of the formal garden. The man of 'nature' says it is unnatural, and it gives an occasion for cheap ridicule too obvious to be resisted. But those who have a weakness for the vicious old practice are in good company. The Romans used to do all sorts of things in pleaching, and so did everybody else down to the end of the seventeenth century and later. The word 'pleach' means the trimming of the small boughs and foliage of trees or bushes to bring them to a regular shape, and, of course, only certain species will submit to this treatment – such as lime, hornbeam, yew, box, holly, white-thorn, and privet, kinds that are 'humble and tonsile', as an old writer calls them.

SIR REGINALD BLOMFIELD, 1892

It must be admitted that one of the great drawbacks to gardening and weeding is the state into which the hands and fingers get. Unfortunately, one's hands belong not only to oneself, but to the family, who do not scruple to tell the gardening amateur that her appearance is 'revolting'. Constant washing and always keeping them smooth and soft by a never-failing use of vaseline – or, still better, a mixture of glycerine and starch, kept ready on the washstand to use after washing and before drying the hands – are the best remedies I know. Old dog-skin or old kid gloves are better for weeding than the so-called gardening gloves; and for many purposes the wash-leather housemaid's glove, sold at any village shop, is invaluable.

MRS C. W. EARLE, 1897

There is great healing power in digging. This is so much the case that one is tempted to wonder if any actual electrical power comes up to one from the earth. Perhaps the benefit is merely from the rhythmic movements of the body. At any rate, however sulking and rebellious one may be at the start, sensitiveness creeps up the fork into hands and body and legs. Finally the brain surrenders and one is again at peace with the garden.

CLARE LEIGHTON, 1935

The moment for mowing has come. We go into the shed and look at our scythe. From the fields around comes the drone of the reaper, sending us almost to sleep with the bee-like hum, now nearer, now further, of its graduated waves of sound. Even the voice of the mower as he calls to his horses has a humming quality. So deeply are we under the spell of this drone that we have hardly the energy to think of our own mowing. The scythe seems heavy as we lift it from its hooks on the shed wall. If that soothing hum would only stop, we could brace ourselves to start. We linger and pause, and look about us, eager for any excuse to delay work. Perhaps the daffodil leaves are not as dead as we had imagined last evening in the half light? But when we go to the orchard to look, we see that there is no excuse for us, they lie crinkled and brown among the roots of the grasses.

The grasses undulate in the breeze, with the motion of a slight swell at sea. As we walk round the orchard, now facing, now backing the sun, they change colour; they are pale silver fawn with the sun full on them, and darker and redder against the sun's light. And as the men and women in a vast crowd have their unremarked, individual beauty and character, moulded each in his own fashion, differing each from the other in shape and colouring, so are the grasses in the orchard composed of multitudinous varying forms, some frail and fine, some erect and sturdy, each with its own pattern of life. I look closer. Where the rapid glance perceives a mere shimmering stretch of fawn, I now see the cock's foot grass, with violet tinted flecks of pollen still sticking to its rough spikes; rye grass and vernal grass are light against brown plantain; the flowered green timothy grass towers smooth and erect, austere of form among the shaking, quivering totter grasses. Shorter, in this vast crowd, clump the 'backbone' grasses, familiar to us from our childhood's game of 'Tinker, Tailor'. Pale meadow soft grass and meadow poa add to the waving buff, a background to some still blossoming pink vetch. Aristocrat amidst this multitude stands the lovely yellow melilot, like a beauty in the market place. Burnt spire tells of seeding dock that we have overlooked in our weedings. Bladder of white campion looks strangely smooth against fringed grasses. At the far end of the grassland grows a white clot of moondaisies. We did not dig the orchard land when we made our garden, merely cultivating the ground that surrounded each fruit tree; so to-day it blossoms with stray lucerne and white clover, heritage from the days when it was pasture.

Beneath these towering, tapering grasses, in the thick tangle of undergrowth, creep little yellow hop trefoil, and scarlet pimpernel. The gilt downy seeds of the goat's beard lie low in the grass, seeking the earth. Below this carpet of flowering weeds are the dark homes of insects. Moss covers the nest of the wild bee, which hums in the short stemmed clover. Butterflies are a brown and blue mist among the grasses. Ripened vetch pods burst in the heat with sudden crackles.

Whether it is the unceasing drone of the reaper, soothing me till all action seems impossible, or whether it is genuine concern for this world of grasses, still I hesitate to mow, telling myself that it is wrong to destroy such beauty.

And then Noel sharpens the scythe. The rasping sound of the whetstone on the blade wakes me from my languor, stinging me like a plunge into icy water. We must start mowing.

CLARE LEIGHTON, 1935

TOOLES & INSTRUMENTS NECESSARY FOR A GARDINER &c.

Three spades of severall sizes
A shovell. 1 Matocck. 1 Pick-Axe
Spade Staff: 3 Hawes of different size
3 Rakes of several sizes & finenesse

A plow-rill for seedes

An Infusing tub hoop'd wth yron

A Water-barrow & Tub: A Cooke-staff

2 Couch forks: 1 Iron crow & spoone

3 Water pots of severall siz'd holes

A Tree-pump

3 sythes, 12 Rubstones, 1 whetstone

1 Grinding stone

An Iron Reele & pin fitted wth whipcord

2 Hand-bills, 1 Hoock-bill & Cycle

2 flower Googes: 3 extracting cases

2 Trowells: 2 paire of shares

1 Water paile: 2 Wooden dibbers

1 yron setting stick

1 paire of wooden compasses

A Measure of 10 foote divided

A Tracing staff shod

2 wheele-barrows

A large levell & one for Uprights

1 Hand-barrow: 3 weeding baskets

1 Sallad-basket

Fruite & Flower baskets of severall sizes

2 weeding knives

1 Wyre seive, one Wyre skreene

1 Fine haire seive

1 Iron roller: 1 stone-Roller

1 Wooden-Roller

2 ladders, a long & shorter

1 Tree stage & ladder to gather fruite on

2 Tressells & a board to remove and clip
    hedges on: or a Forme to raise

1 steele vermine trap

3 Mouse Traps, Moule Traps, grained
    to strike the heaving moule

1 bird clapper: 1 Trap-cage

A Net to preserve seedes from Birds

1 Tray to clense & van seedes

Saile-cloths to dry seedes on and hang
before blossoming wall fruit

12 Mattrasses to cover Beds with

A case of drawers for seedes

A Tin-box with divisions for seedes: and
flower rootes

Bags & paper hoods for seedes

Flower shades, & for newly set plants

Wooden squares, & one well glaz'd for
the Hott-beds

50 Melon glasses

1 Turfe beater: 1 Rammer

1 Measuring chaine

1 Halfe-rod-pole

1 Water-levell

2 pruning knives

2 Graffing [grafting] knives & Iron Wedge

1 Inoculating Gauge

2 Graffing saws: 1 Hatchett

1 Broad strong chezell

1 Mallet: 1 Augur: 2 Gimletts

1 Shaving-knife

2 Hammers

1 Naile box divided in 4

1 paire of pincers

1 paire of pliers for Wyre

2 files: 1 Betell

1 Chopping block

A set of Letters, & Figures

A Bushell, Gallon, pint measure

A parcell of Wyres

A basket of shreads & felt cuttings: to tack

        up wall fruite
    Bundle of ash-poles
    Stakes
    Tubs, cases & Boxes, Flower pots
    Layer pots, shading potts, of severall sizes
    Canes, Hoofes, Insect glasses
    Pack threid & Basse to bind with
    Saw-dust, sand, Mosse
    A Ruler, Black lead pen: paper
    Behives of all sizes
    Tallies of Lead
    Finer threid to bind up Nose-gays wth
    A paper book to note what when &
        where he sows & plants, & register
        the successe of tryalls
    The Bee Calendar

                    JOHN EVELYN, 1687 (pub. 1932)

A practical attention to a garden, by many, is esteemed degrading. It is true, that pastoral and agricultural manners, if we may believe the dignified descriptions of Virgil, are greatly degenerated. The employments of shepherds and husbandmen are now mean and sordid. The care of the garden is left to a peasant. Nor is it unreasonable to assign the work, which wearies without amusement, to those, who are sufficiently amused by the prospect of their wages. But the operations of grafting, of inoculating, of pruning, of transplanting, are curious experiments in natural philosophy; and, that they are pleasing as well as curious, those can testify, who remember what they felt on seeing their attempts succeed . . .

It is the lot of few to possess territories like [William Shenstone's], sufficiently extensive to constitute an ornamented farm. Still fewer are capable of supporting the expence of preserving it in good condition. But let not the rich suppose they have appropriated the pleasures of a garden. The possessor of an acre, or a smaller portion, may receive a real pleasure, from observing the progress of vegetation, even in a culinary plant. A very limited tract, properly attended to, will furnish ample employment for an individual. Nor let it be thought a mean care; for the same hand that raised the cedar, formed the hyssop on the wall. Even the orchard, cultivated solely for advantage, exhibits beauties unequalled in the shrubbery; nor can the green-house produce an appearance to match the blossom of the apple and the almond.

VICESIMUS KNOX, *c.* 1778

Many curious and interesting words connected with gardening are to be heard along the Welsh border, that delightful stretch of country lying between the prosperous central plain of England and the 'gaunt purple steeps of Wales'. One must adventure into the byways on either side of the serpentine Severn to make these philological discoveries which well repay the trouble taken.

The names the borderer gives to his garden tools would puzzle many a gardener living farther afield. He would almost certainly fail to recognize a grafting shovel as a digging spade, neither would he know a spittle as the same tool, nor a spittle tree as a spade handle. To the border counties gardener a garden fork is a yelve, dungevil, sharevil, or sherevil, while his Dutch hoe he calls his scuffle, since he scuffles weeds rather than hoes them. A trindle is the wheel, and gudgeons the iron pivots in the wooden axle of his wheelbarrow.

Along the border seedness is the general term for seedtime,

lent-sidness is spring seedtime. If a garden is dry, loose, and crumbling it is said to be mildy. The term Pea sticks is never heard in the native vocabulary for these are Pea risers or rises, and on occasion stickings. If the weather is what is generally called close growing weather then in the border counties it is muckery.

If in a drought plants flag and then revive on the coming of rain they are said to prink up, and should they be laid flat by rain or wind they will be described as lodged. A hit is a good crop which is often described as profligate rather than prolific, though by dictionary definition this is hardly fair to the individual innocence of the plants. Sometimes a crop, as for instance of Mushrooms, may be called a flight, while the word mort means an abundance. In digging, a graff is one spade's depth, while to delve is to go one spade's depth farther. The borderer refers to a small quantity of manure – even up to a load – as a jag of muck. Land which has been impoverished by lack of manure is said to be beggared – literally true of course.

The rather strange word ackerspire means to sprout Potatoes, and if the same Potatoes unluckily get cut by frost they are said to be frost ketchen. Should rain be hindering and holding up gardening operations it is defined as lattin. The border gardener never allows a tellif to develop in his garden, for if he did so he would have a thick tangled crop of weeds to contend with. Any plant that gets uprooted is said to be root wouted, and a ronk plant is one vigorous in growth. The gardener's sweepings up are called mullock or rouse, his pease haulms are peasum, and his Potato tops wisles.

A plot of ground is always known as a but, though the wife of the cottager may refer to her flower plot as her posy knot. A heyment or hayment is a fence or boundary, and should this be a hedge that has a broken-down gap in it the gap will be referred to as a glat. The spreading roots of trees are known as spurns, while an imp is a scion or shoot. If the land is difficult to work owing to wet patches it is said to be saky.

It would be odd if this border nomenclature did not extend to the plants, so it is no surprise to find Shallots called sheelots or scallions, and Onions defined as inions by the alteration of the initial vowel.

Tulips are known as quilps, while the common Lilac, *Syringa vulgaris*, is called laylock. The Sunflower is known as a wall-flower which may explain the rhyme, 'Wall-flowers, wall-flowers, growing up so high'. The Evening Primrose is called the Naked Lady, maybe from the fact that she discards some part of her clothing every night.

When Cabbages remain open and slow to heart they are termed jowters and very small Potatoes are known as rattocks or chats. A poor quality, rough-tasting Apple is known as a koling, and the dark purple Plum which ripens the first week in August, namely Lammas-tide, is known as a lammas or lammas plum.

Scutch or squitch are fairly general names for Couch-grass, so this use along the border is not exceptional, but surely the describing of the small Bindweed as Devil's guts shows in what slight respect it is held. Sometimes this pernicious weed is more politely called Billy Clipper.

Here is a word of warning. Should you be eavesdropping on a border gardener and hear him refer to his marrow don't at once conclude that he is referring to some succulent monster on his Marrow bed, for he may be speaking of a mate, a fellow workman, or a friend, for a marrow means any one of these, and a marrow to him is not always encountered in the garden, the encounter may take place in the local.

H. CLAYTON JONES, 1951

In the summer of 1829, Mr N. B. Ward, of Wellclose Square, London, noticed a seedling of the common Male Fern and a seedling of one of the commonest British Grasses (*Poa annua*) growing on the surface of some moist mould in a large bottle. Curious to observe how vegetation would proceed in so confined a situation, he placed the bottle, loosely covered with a tin lid, outside one of his windows, with a northern aspect. This cover allowed a sufficient change of air for the preservation and development of the plants, and, at the same time, prevented the evaporation of the moisture within. These plants remained in the bottle for more than three years, during which time not one drop of water was given to them, nor was the cover removed. The Grass flowered the second year, but did not ripen its seeds. The Fern developed five or six fronds annually, but did not produce any fructification.

This led Mr Ward to experiment on the growth of plants in closed cases, and to advocate the use of glazed boxes for bringing home plants from all parts of the world, and also for growing many kinds of plants in rooms, which could not otherwise be grown in such an atmosphere. Thus every description of closed glass case used for growing plants, from a simple bell-glass to the most elaborate mini-ature green-house, came to be designated a Wardian case.

WILLIAM THOMSON, *c.* 1890

The common watering pot for the Garden beds with us, hath a narrow neck, big belly, somewhat large bottom, and full of little holes, with a proper hole formed on the head to take in the water, which filled full, and the thumb laid on the hole to keep in the aire, may on such wise be carryed in handsome manner to those places by a better help aiding, in the turning and bearing upright of the bottom of this pot, which needfully require watering.

. . . the beds at one instant shal not fully be watered, but as the earth and plants drink in, so gently sprinkle forth the water, in feeding the plants with moisture, as by a brest or nourishing Pap, which like handled, shall greatly prosper the tender plants comming up, where they otherwise by the hasty drowning with water, are much annoyed, and put in hazard of perishing.

THOMAS HILL, 1577

One of the most teasing things that a lady can meet with in the cultivation of her plants, is a badly-made watering-pot, at least such is my own experience. To be sure, my blue apron is not much the worse for a daily sprinkling of water from leaky pots or spouts, and I seldom use gloves when I am watering; but ladies, whose clothes are not suited to resist wet, must be sadly annoyed when at the delightful exercise of watering their own favourite flowers, if their watering-pots let off a single drop of water except by the spout or rose. Indeed, even watering with an open-spout pot is not fit work for ladies; and as they should always use the rose, the danger of leakage is increased in their case when there is any sediment in the water, as it soon fills up the small holes in the rose, and in order to get this cleared from time to time, the rose is made to come off. After a little use the water finds a way of escaping at the junction between the detached rose and the spout of the watering-pot, and a stream of water is sure to descend on the shoes; or, if the plants should be on a high shelf, down the sleeves, or over the clothes. I have made a slight improvement on watering-pots to do away with these annoyances; and, I dare say, it will be useful to many lady gardeners. It is to have the rose fixed, instead of coming off in the usual way, and to have a *moveable* square piece of tin inside the watering-pot to cover the hole at the bottom of the spout, or where it enters the pot. This piece of tin must be pierced with small holes, the same size as those in the rose, and this will so far filter the water that nothing can reach the rose, but what can pass through it. When the piece of tin becomes choked up, it may be taken out to be cleared; and when in use it is kept in its place by two strips of the same material, soldered to the side of the watering-pot. Any tinsmith can add this strainer in five minutes when he is making the watering-pot, if he is told that a strainer is wanted inside the watering-pot over the spout to slide up and down like a carriage window.

MR BEATON (gardener to Sir William Middleton), 1841

Good gardeners tell you never to cut flowers except with a sharp knife. This is good advice for shrubs or pot-plants, the clean cut being better for the plants; but I advise that the knife should be on a steel chain a foot or so long, with a good pair of garden hook-shaped scissors at the other end – for the cutting of annuals or lately planted plants with a knife, in light soil, is very much to be avoided. The smallest pull loosens the roots, and immediate death, in hot weather, is the result. Another advantage of knife and scissors together on the chain is that they are more easy to find when mislaid, or lost in the warm and bushy heart of some plant.

MRS C. W. EARLE, 1897

When I was travelling on the hills of Hong-Kong, a few days after my first arrival in China, I met with a most curious dwarf *Lycopodium*, which I dug up and carried down to Messrs Dent's garden, where my other plants were at the time. 'Hai-yah,' said the old compradore, when he saw it, and was quite in raptures of delight. All the other coolies and servants gathered round the basket to admire this curious little plant. I had not seen them evince so much gratification since I showed them the 'old man Cactus' (*Cereus senilis*), which I took out from England, and presented to a Chinese nurseryman at Canton. On asking them why they prized the Lycopodium so much, they replied, in Canton English, '*Oh, he too muchia handsome; he grow only a leete and a leete every year; and suppose he be one hundred year oula, he only so high,*' holding up their hands an inch or two higher than the plant. This little plant is really very pretty, and often naturally takes the very form of a dwarf tree in miniature, which is doubtless the reason of its being such a favourite with the Chinese.

The dwarfed trees of the Chinese and Japanese have been noticed by every author who has written upon these countries, and all have attempted to give some description of the method by which the effect is produced. The process is in reality a very simple one, and is based upon one of the commonest principles of vegetable physiology. We all know that any thing which retards in any way the free circulation of the sap, also prevents to a certain extent the formation of wood and leaves. This may be done by grafting, by confining the roots, withholding water, bending the branches, or in a hundred other ways which all proceed upon the same principle. This principle is perfectly understood by the Chinese, and they make nature subservient to this particular whim of theirs. We are told that the first part of the process is to select the very smallest seeds from the smallest plants, which is not at all unlikely, but I cannot speak to the fact from my own observation. I have, however, often seen Chinese gardeners selecting suckers and plants for this purpose from the other plants which were growing in their garden. Stunted varieties were generally chosen, particularly if they had the side branches opposite or regular, for much depends upon this; a one-sided dwarf tree is of no value in the eyes of the Chinese. The main stem was then in most cases twisted in a zigzag form, which process checked the flow of the sap, and at the same time encouraged the production of side branches at those parts of the stem where they were most desired. When these suckers had formed roots in the open ground, or kind of nursery where they were planted, they were looked over and the best taken up for potting. The same principles, which I have already noticed, were still kept in view, the pots used being narrow and shallow, so that they held but a small quantity of soil compared with the wants of the plants, and no more water being given than what was barely sufficient to keep them alive. Whilst the branches were forming, they were tied down and twisted in various ways; the points of the leaders and strong growing ones were generally nipped out, and every means were taken to discourage the production of young shoots which were possessed of any degree of vigour. Nature generally struggles against this treatment for a while, until her powers seem in a great measure exhausted, when she quietly yields to the power of art. The Chinese gardener, however, must be ever on the watch, for should the roots of his plants get through the pots into the ground, or happen to be liberally supplied with moisture, or should the young shoots be allowed to grow in their natural position for a short time, the vigour of the plant which has so long been lost will be restored, and the fairest specimen of Chinese dwarfing destroyed. Sometimes, as in the case of peach and plum trees, which are often dwarfed, the plants are thrown into a flowering state, and then, as they flower freely year after year, they have little inclination to make vigorous growth. The plants generally used in dwarfing are pines, junipers, cypresses, bamboos, peach and plum trees, and a species of small-leaved elm.

ROBERT FORTUNE, 1847

As touching remedies against the Frogges, which in Summer nights are wont to be disquieters to the wearied Husbandmen, through their daily labour, by chirping and loud noise making, let the Husbandman exercise this helpe or secret, borrowed of the skilful Greek *Africanus*, which is on this wise: Set on some bank (saith he) a Lanthorne lighted, or other bright light before them, or on some tree (fast by) so hang a light, that by the brightness of the same light, it may so shine upon them, as it were the Sunne, which handled on this wise, wil after cause them to leave their chirping and loud noise making.

THOMAS HILL, 1577

The worst ENEMYES to gardens are Moles, Catts, Earewiggs, Snailes and Mice, and they must bee carefully destroyed, or all your labor all the year long is lost.

Many storyes there are of pretty wayes to catch Moles, as by putting a pott in the earth with a live Mole in it, to which the rest will resort, and falling in cannot get out againe; by pouring of scalding or poyson'd waters upon their great hills where their nests are and constant dwellings; and diverse others; but I passe them over as fabulous or very little usefull. The only assur'd meanes to destroy them is by watching them heave at sun rising & setting, and then casting them forth with a spade, or striking them with a Mole speare, but I leave this to the direction of the Mole catchers who are everywhere to bee found, and shall rather say somthing of preventing the mischiefes these creatures doe by well walling-in your ground, for if the foundation be anything deepe they cannot get under, for they ever run neare the face of the earth when they are underground, and if the doores shut close at the bottomes, and there bee no holes in the walls, they cannot possibly enter above ground, for they run not up precipices as ratts or mice.

As for Catts they doe much hurt in most places, espetially in Townes if not prevented, for when the earth hath beene lately digged up, as it must bee newly before the planting of your flowers, they delight to scrape in it, and urine and dung upon it. To help which there is no better way than to cover the new planted beds with netts fastned downe close upon them with pinns of wood, which must not bee removed till the earth bee well sadded, and then they are in no great danger.

Earewiggs hurt most Gilliflowers, and are taken best when they are newly podded (for they feed upon the yong pods most) with sheepes hooves stucke upon stickes by the flowers, into which they creepe in the morning to hide themselves all day, feeding all night, and then you shall bee sure of them every morning and may easily kill them.

Snayles doe the leaste harme, and may bee taken in the night in the sommer with a candle as they creepe about, or early in the morning, or after raine.

As for Mice, many grounds are not troubled with them, and where they are they are easily taken with little traps baited with nut kernells, pieces of apples and such things. They are only field mice that trouble the garden, and they hurt most Crocus rootes, being the tenderest and sweetest.

Thus have wee past through some of the chiefest Generall things that were fittest to be knowne before we come to Particulars, and shall say no more, but that all Flowers love to lye soft and dry, soe plant them in due season, and let the beds bee of fine earth, and lye convex, that is somewhat high in the middle, to cast off wett, and let them bee well weeded and kept from the Enemyes, violent weathers, either hot or cold or wyndy, and hurtfull creatures, and your expectation shall bee crown'd with pleasure and delight; for by good cultivation and defence only you shall arrive at the greatest perfection flowers are capable of; for there is no knowne Art to give them such colours as wee desire by steeping the rootes or seedes in colour'd waters, or by putting any ingredients into them that shall worke that effect, nor any way by observing the Moone or heavens to make flowers larger or more double, or to worke such wonders as are both sayd and written to amuse and deceave the unexperienced and credulous.

SIR THOMAS HANMER, 1659

A large body of the army of the small ones of the earth has attacked us, and it is no fault of theirs if we are not despoiled of the best of our spring delights. The field-mice have at length found out the Crocuses; we, on our side, have set traps in their way, and large numbers have fallen – quite flat, poor little things – under the heavy bricks. We believe we should have slain many more, had not some clever creature made a practice of examining the traps during the night, devouring the cheese, and in some way withdrawing the bit of stick, so as to let the brick fall harmless. Suspicion points towards one person especially – the old white fox-terrier, who lives in the stables, and is master (in his own opinion) of all that department, and whom neither gates nor bars can prevent going anywhere he chooses to go. – 'Impossible!' says he with Mirabeau, 'don't mention that stupid word!' Up to this time field-mice have not troubled us much. In the days when there was always a hawk or two hovering over the ploughed land, or keeping watch over the green meadows, and when we used to hear the owls in the summer nights, and saw the white owl who lived somewhere near by, sail silently in the grey of evening across the lawn – in those days we knew little of the plague of field-mice. But now we have changed all that; cheap gun licences have put a gun into every one's hand, the vermin is ruthlessly shot, and the balance of Nature is destroyed.

MRS E. V. BOYLE, 1883 (pub. 1884)

We have a cat, a magnificent animal, of the sex which votes (but not a pole-cat), – so large and powerful, that, if he were in the army, he would be called Long Tom. He is a cat of fine disposition, the most irreproachable morals I ever saw thrown away in a cat, and a splendid hunter. He spends his nights, not in social dissipation, but in gathering in rats, mice, flying-squirrels, and also birds. When he first brought me a bird, I told him that it was wrong, and tried to convince him, while he was eating it, that he was doing wrong; for he is a reasonable cat, and understands pretty much everything except the binomial theorem and the time down the cycloidal arc. But with no effect. The killing of birds went on to my great regret and shame.

The other day I went to my garden to get a mess of peas. I had seen, the day before, that they were just ready to pick. How I had lined the ground, planted, hoed, bushed them! The bushes were very fine – seven feet high, and of good wood. How I had delighted in the grow-ing, the blowing, the podding! What a touching thought it was that they had all podded for me! When I went to pick them, I found the pods all split open, and the peas gone. The dear little birds, who are so fond of the strawberries, had eaten them all. Perhaps there were left as many as I planted: I did not count them. I made a rapid estimate of the cost of the seed, the interest of the ground, the price of labour, the value of the bushes, the anxiety of weeks of watchfulness. I looked about me on the face of Nature. The wind blew from the south so soft and treacherous! A thrush sang in the woods so deceit-fully! All Nature seemed fair. But who was to give me back my peas? The fowls of the air have peas; but what has man?

I went into the house. I called Calvin (that is the name of our cat, given him on account of his gravity, morality, and uprightness. We never familiarly call him John). I petted Calvin. I lavished upon him an enthusiastic fondness. I told him that he had no fault; that the one action that I had called a vice was an heroic exhibition of regard for my interests. I bade him go and do likewise continually. I now saw how much better instinct is than mere unguided reason. Calvin knew. If he had put his opinion into English (instead of his native catalogue), it would have been: 'You need not teach your grandmother to suck eggs.' It was only the round of Nature. The worms eat a noxious something in the ground. The birds eat the worms. Calvin eats the birds. We eat – no, we do not eat Calvin.

CHARLES DUDLEY WARNER, 1876

. . . the relatively simple one-sided arrangement which is so prevalent in the plant world is difficult enough to understand. Geology seems to demonstrate that the earliest flowering plants depended, as the conifers do today, upon the chance that some of their abundant pollen would be carried by the wind to the waiting ovaries. Then, since all organic matter is potentially edible by something, it is assumed that certain insects got into the habit of eating pollen, accidentally got some of it entangled in the hair on their bodies as many still do, and accidentally rubbed some of it off on the stigmas of the other flowers they visited. Since, for the plant, this was more effective than wind pollination and involved less waste of vital material, those plants which were most attractive to insects got along best. And as the degree of attractiveness accidentally varied, 'natural selection' favored those which were most attractive, until gradually all the devices by which plants lure insects or birds – bright colored petals, nectar which serves the plant in no direct way, and perfume which leads the insect to the blossom; even the 'guide lines' which sometimes mark the route to the nectar glands – were mechanically and necessarily developed.

Gardeners usually hate 'bugs', but if the evolutionists are right, there never would have been any flowers if it had not been for these same bugs. The flowers never waste their sweetness on the desert air or, for that matter, on the jungle air. In fact, they waste it only when nobody except a human being is there to smell it. It is for the bugs and for a few birds, not for men, that they dye their petals or waft their scents. And it is lucky for us that we either happen to like or have become 'conditioned' to liking the colors and the odors which most insects and some birds like also. What a calamity for us if insects had been color blind, as all mammals below the primates are! Or if, worse yet, we had had our present taste in smells while all the insects preferred, as a few of them do, that odor of rotten meat which certain flowers dependent upon them abundantly provide. Would we ever have been able to discover thoughts too deep for tears in a gray flower which exhaled a terrific stench? Or would we have learned by now to consider it exquisite?

JOSEPH WOOD KRUTCH, 1954

Of all the creatures which are behoveful for the use of man, there is nothing more necessary, wholsome or, more profitable than the Bee; nor any less troublesome, or less chargeable ... It is a creature gentle, loving and familiar about the man, which hath the ordering of them, so he comes neat, sweet and cleanly amongst them, otherwise if he have strong and ill smelling savours about him, they are curst and malicious, and will sting spitefully, they are exceeding industrious and much given to labour, they have a kind of government amongst themselves, as it were a well ordered Common-wealth, every-one obeying and following their King or Commander, whose voice (if you lay your ear to the hive) you shall distinguish from the rest, being louder and greater, and beating with a more solemn measure. They delight to live among the sweetest herbs, and flowers, that may be, especially Fennel, and Wall Gilly flowers, and therefore their best dwellings are in Gardens; and in these Gardens, or near adjoyning thereunto, would be divers fruit-trees growing, chiefly Plum-trees, or Peach-trees; in which when they cast, they may knit without taking any far flight, or wandring to find out their rest. This Garden also would be well fenced, that no Swine nor other Cattel may come therein, as well for overthrowing their Hives, as also for offending them with their ill savours. They are also very tender, and may by no means indure any cold; wherefore you must have a great respect to have their houses exceeding warm, close, and tight, both to keep out the frost and snow, as also the wet and rain; which if it once enter the Hive, it is a present destruction.

GERVASE MARKHAM, *c.* 1638

With tumbled hair of swarms of bees,
And flower-robes dancing in the breeze,
With sweet, unsteady lotus-glances,
Intoxicated, Spring advances.

From an anthology of Sanskit poems of the 12th–15th centuries,
(trans. JOHN BROUGH, 1968)

Spring is the most skilful of all gardeners, covering the whole ground with flowers, and shading off the crudest contrasts into perfect harmony; and were it April, May, and June all the year round, I, for one, would never again put spade or seed into the ground. I should select for the site of my home the heart of an English forest, and my cottage should stand half-way up an umbrageous slope that over-looked a wooded vale, from which majestic trees and coverts again rose gradually up to the horizon. One would make just clearance enough to satisfy one's desire for self-assertion against Nature, and then she should be allowed to do the rest. What are all the tulips of the Low Countries in point of beauty compared with the covering and carpeting of the wild-wood celandine? Your cultivated Globe-Flower and Shepherd's-Bane are well enough; but they have a poverty-stricken look when paragoned with the opulent splendour of the marsh-marigold, that would then grow along the moist banks of the low-lying runnels of my natural garden.

Perhaps I should be accused of exaggeration were I to describe the effect produced on my, no doubt, not impartial gaze by the *Anemone apennina* and the *Anemone fulgens* now in full bloom in the Garden that I love. Professional gardeners will tell you, in their offhand way, that these will grow anywhere. They will not – being, notwithstanding their hardiness in places that are suitable, singularly fastidious as to soil and situation, and even sometimes unaccountably whimsical in our un-certain climate. The *Anemone fulgens*, or *Shining Windflower*, is common enough, no doubt, where it chooses to thrive, and you may see it in bloom in open and favourable Springs as early as the month of Febru-ary, while, with proper arrangement of aspect, you can prolong its dazzling beauty well into May. But the *Anemone apennina*, which I have known some people call *The Stork's-Bill Windflower*, is, as far as my experience goes, rarely seen in English gardens. It used, an indefinite number of years ago, to be sold in big basketsful by dark-eyed, dark-haired, dark-skinned flower-girls in the Via Condotti in Rome, in the months of February and March, and I recollect a good Samaritan putting the finishing touch to my convalescence, after a visitation of Roman fever, by bringing to my room a large posy of this exquisite flower, varying in colour from sky-blue to pure white, and springing out of the daintiest, most feathery foliage imaginable. Perhaps, therefore, it is in some degree the spell of association which makes me feel tenderly enthusiastic concerning the Apennine Windflower. I do not say it prospers in our latitudes as it does in the sunshine-shadow of the Appian Way. But, in most years, it maintains itself against rude winds, unkindly leaden clouds,

> And Amazonian March with breast half bare,
> And sleety arrows whistling through the air.

It asks for some but not too much shelter, and I have had to lighten the natural heaviness of my ground, in order to humour it, with well-pulverized soil and a judicious contribution of sand.

But, with all my partiality for these domesticated windflowers, I will not pretend that they can hold a feather to undulating stretches of sylvan anemones; and in April these would be as numerous as the pink-and-white shells of the sea-shore, which in colour they curiously resemble, around my forest abode. Blending with them in the most affable manner would be the wild or dog-violets, destitute of scent, but making amends by their sweet simplicity for the ostensible absence of fragrance. Where they rule the woodland territory the earth is bluer than the sky.

ALFRED AUSTIN, 1894

June is the month that takes care of itself. Even the dullest garden can't help being colourful in June. When the cow parsley reaches shoulder level in the hedgerows and the roadside is scented with honeysuckle and wild roses the garden too seems to grow up overnight. This is the time when one discovers if one has planted too closely, and I always have, and if one has staked sufficiently and efficiently, and I never have.

June is the month when roses tumble over the walls, the tall spikes of delphiniums tower above the jungle of the borders, at the mercy of the gales that nearly always turn up some time in June, to humble our pride and challenge our foresight. The farmers take the 'June drop' in their stride. Though it flattens the corn and brings violent rain to devastate the hay, it also thins out the apples for them. The poor gardener has no such compensations for his shattered hopes.

MARGERY FISH, 1965

This morning I woke to autumn – a warm autumn morning rather than the cool summer morning of yesterday. It's like the difference between a fiddle and a 'cello both playing middle C: the same note, but another quality. There was a light veil of mist on the meadow, dew heavy on the rough grass where the colchicums are in flower; and nearby in the shorter grass the first autumn crocuses were showing. I grow a bit bored with my garden as August ebbs, but the colchicums and crocuses always rouse me. The waning of the year is melancholy, but the crocuses are like spring at one's feet.

WILFRID BLUNT, 1963

Autumn adds such wonderful touches of happy accident that, when it comes, really comes, a wise man leaves his garden alone and allows it to fade, and wane, and slowly, pathetically, pass away, without any effort to hinder or conceal the decay. Indeed, it would be worth while having a cultivated garden if only to see what Autumn does with it.

ALFRED AUSTIN, 1894

Jorrocks in autumn was all cock-a-hoop when the first considerable frost told him the hunting season had begun in earnest. 'Blister my kidneys!' he cried. 'The dahlias are dead!'

TYLER WHITTLE, 1969

In September, leaves begin to thin and make space. Even the days shorten and ripen.

Every clear morning grass and leaves bead with water, afternoons end sooner, and evenings carry a cold edge; shady places stay damp all day. Cloudless nights bring frost to the hill, and its bracken turns yellow further down the slopes each week. Birch leaves grow yellow, too, falling lightly on the paths, so that brown and amber trails wind through the green.

Summers can be unbelievably extended, with autumn two crippling nights that shred unhardened buds and stems; but this season is best when it approaches gently without sharp frost or wind, and takes its time to dismantle summer, tree by tree. Then, warm September sunlight throws pink shadows about wild blue scabious, and Small Tortoiseshell butterflies drink their last luxury; but for the chill in the grass, it could be a July afternoon. The sun slants lower each day between trees, illuminating areas shaded all summer, and the canopy lightens; a new brightness enters woodland. Colour comes back, as tired August greens sharpen to yellows and orange, washed lucent by rain.

Autumn can be our brightest period. Everywhere, from deep in the Gorge up to morning-frosted slopes of turf and heather, and everything, stem, leaf, berry and bud, wakens to a carnival of colour: yellow, orange, reds and purples glowing under long shadowed sunlight and violet clouds. Flowers become superfluous, not just to a restful garden, but to a brilliant one.

Most flowers are relics, flinching to brown as they open or scowling above dislocation. The herbaceous community is sad enough now, a last chapter in Proust. Fuchsia, an exception, enters more graciously into the geriatrics of autumn, scarlet and purple as ever, white dust of anthers still precise; as autumn withdraws about them, these bushes – urbane in formal surroundings, at ease in the wild – remain confident in their own extent of sun, comforting to have around the house; until the first considerable frost breaks and discards them.

This is also the time of gentians, smokily Plutonic as Lawrence could desire, but even these are survivors, torches going down, an interruption of the great slow swing towards winter. Autumn crocus rhymes better with the season, saying all it needs in naked unabashed, leafless, flower, then dying down. Not strayed from summer, it presages a change to austerity, and its simple bloom sends you ahead to snowdrops and the first clean blossoming of spring.

I must briefly sketch our autumn colours, though they cannot rival the more varied panoply of wealthier places. Our own incendiaries burn against silver-lichened boulders and the smouldering tones of surrounding hillsides; they ignite from acid-yellow moss through russet bracken to the fuming crimson of tall red oaks, and all display a simple zest, everyone sharing in the feast of leaf and berry.

So you are to imagine high skies of light blue, travelled by pale cirrus and washed by showers of cold rain. To the north clouds pile black and purple; from early October, hills are dusted white against them. Sounds echo crisp through the tang of autumn, that long after-taste of summer. And so do silences . . .

Birch is still our framework, its gold swimming above butter-yellow hazel, that yellow slapped on flat like butterpats, a still life beneath the flicker of birch. Below, too, is beech; all those hedges and miniature groves switch on their amber, orange and even purple and crimson at different rates, some yet green when others have run down to the last wet chestnut, a luminous harmony behind moss sprinkled with white, pink, olive and every whisper of red and brown. Beech, as dwarf woodland or intermittent bonfires beneath high conifer, is necessary as birch and hazel to autumn.

Outstanding also are blaeberry, running ruby and orange drifts between coal-black junipers and leathery rhododendrons; barberries, some an indescribable signal-red, orange-red berries enamelling their arching wands; and azaleas – with yellow, orange, scarlets, ruby-crimson and purples, darkening to deep blue, chasing across them, and central in each glaring rosette gleams next year's apple-green bud.

The lowest indispensable storey is bracken: an ochre, bleaching to cream, deepening to squirrel brown beneath the buttery hazels. When all leaves are down it dries to a foxy fur and loses its architecture; but that no longer matters, for the bones of the garden are visible again and bracken is free to add informal texture, across which pheasants saunter and crackle their own autumnal circumstance. Tiresome other times, these birds from their splendour this season find sanctuary here from ritual disturbance further down the glen . . .

Many fine colourists of spring repeat their excellence in autumn. Larch opened shrill green and now closes with equally startling yellow, sharpening to bright orange, flattening to old gold, and falling away then softly; or vanishing altogether in one night of wind. An early snowfall drips glassily from its lime-green needles. Aspen, then liquid silver, now coins doubloons of real gold, a gilding distinct from orange and brown either side or any counterfeits of terracotta. Aspen gold here is up to Colorado standard, and from pale blue sky falls leisurely in handfuls, strewing the paths with an ample gesture of sovereigns. You kick richly through them.

I must dismiss our Sorbus species [*Sorbus vilmorinii*] – whose finest season this is – in a single sentence reserved for one kind laden with ox-blood berries that pale to rose and then to bright almond-pink, while neat leaflets flush red, maroon, purple and finally bronze before falling and leaving the naked tree, bowed beneath its fruit, as astonishingly luminous in November (despite the fieldfares) as a flowering almond in early spring.

In autumn, evergreen conifers step forward again to reassure you against winter. Those with a glaucous edge of silver about their needles best set off deciduous yellows and reds – Sitka spruce or Scots pine, for example, preferable to the greeny Norway spruce or lodgepole pine when fronting ripening hill-grass and blue distances. Fiercely pungent Sitka spruce, towering waxy blue behind red oak in tawny evening light, are memorable. Without conifers autumn would lose much of its excitement: just to walk under blue spruce and pine after leaving the orangey gold of larch and aspen is to taste the very air smoky, so overwhelming is the transformation about you.

In late autumn the air often smokes with rain, pelting the last leaves off trees and flooding the Burn, which rises fast. Heard from the house its rumble breaks to a bellow as you open the door. Beside it, your spine rattles to the thud and grind of travelling boulders and you smell that 'unforgettable, unforgotten' racing white water. No rock, cascade or pool is visible, just a huge rope belabouring the Gorge. At the garden bridge it is thirty-five feet across, pier to pier, its walloping fringe snuffling last summer's stems and blowing Guinness froth about your feet. When highest it makes least noise – a swift alarming *swoosh* – so much is then above boulders and bankside rocks; convex, bulging along the middle, a terrifying muscular punch. Dippers flit complainingly, but prudently, about the shaking banks.

Yet below all this, Himalayan primulas [*Primula florindae*] survive, reasserting themselves in spring through the boulders and shingle yearly redistributed around them. Tons of mica-schist are bullied downstream, but primulas preside over their summer in exactly the same place.

After a spate, scenery has changed: fragile waterfalls abolished, cliffs dissolved to rapids. The lip of the big waterfall has retired upstream ten feet in thirty-five years, and a tower beside it likewise; summit hazels straddle with bare roots, and a sprucelet I pushed into its flat blaeberry top all those years ago is now a tall tree whose time has come, a foot from the plunging slab. A favourite deep-diving pool beneath is choked with fangs of rock, monstrous sharp-angled fragments scarred by impact, the wounded cliff peering from above; occasionally a birch stands on its head in the water, loose boulders collected on up-ended roots. Dynamic gardening, beside the Burn.

G. F. DUTTON, 1995

THE BURNING OF THE LEAVES

Now is the time for the burning of the leaves.
They go to the fire; the nostril pricks with smoke
Wandering slowly into a weeping mist.
Brittle and blotched, ragged and rotten sheaves!
A flame seizes the smouldering ruin and bites
On stubborn stalks that crackle as they resist.

The last hollyhock's fallen tower is dust;
All the spices of June are a bitter reek,
All the extravagant riches spent and mean.

All burns! The reddest rose is a ghost;
Sparks whirl up, to expire in the mist: the wild
Fingers of fire are making corruption clean.

Now is the time for stripping the spirit bare,
Time for the burning of days ended and done,
Idle solace of things that have gone before:
Rootless hopes and fruitless desire are there;
Let them go to the fire, with never a look behind.
The world that was ours is a world that is ours no more.

They will come again, the leaf and the flower, to arise
From squalor of rottenness into the old splendour,
And magical scents to a wondering memory bring;
The same glory, to shine upon different eyes.
Earth cares for her own ruins, naught for ours.
Nothing is certain, only the certain spring.

LAURENCE BINYON, 1944

Today, mid-December, the light in the Cotswolds is electrifying, brilliant. The air is cold and crisp, the sky a reflecting blue, the earth black, the leaves and the sky motionless, the clouds a thin haze on the horizon. The only movement to be seen is the swift flight of the finches, tits, sparrows and the occasional wren as they search for food in bushes and borders. Wherever I look, there is a feeling of repose and happiness among the plants . . .

   Then there are those glorious days when the ground is covered with snow. I admit I love snow particularly for the moments when, with a clear conscience, I can stay indoors and watch from inside – it is beautiful even when the snow is falling thickly and I am seeing the garden through a veil.

The evening before, the clouds will have started to look heavy, and it is no surprise next morning to wake to that familiar white light bouncing off the ceiling, a reflection of the snow carpet outside. Look out and all is clean and pure. If the fall has been light, the shrubs and trees stand out crisply. As the sun appears the shadows have clear, strong outlines. There is a muffled silence; familiar sounds are absorbed, distant noises come closer. For the first few hours no human footprints scuff the surface, but there are the tell-tale marks of a rabbit or a hare or sometimes a fox taking a short cut through the garden in search of breakfast. I try hard to persuade everyone to keep off the unmarked snow.

If the fall has been heavy, the contours of the garden are transformed. Each shrub is quite enveloped, each tree sagging with its unaccustomed burden. This protective blanket safeguards precious plants from sharp attacks of frost and cold; far better that they should stay covered than that the snow should melt too quickly, leaving the sap to freeze in the stems. But this advantage is secondary to the beauty of the garden under the snow. I like to think of Vita Sackville-West planting her White Garden at Sissinghurst 'under the first flakes of snow'.

ROSEMARY VEREY, 1988

We are now in the depths of winter ... my first winter at the cottage ... and the first winter when I went mad.

The average gardener, in the cold dark days of December and January, sits by his fire, turning over the pages of seed catalogues, wondering what he shall sow for the spring. If he goes out in his garden at all it is only for the sake of exercise. He puts on a coat, stamps up and down the frozen paths, hardly deigns to glance at the black empty beds, turns in again. Perhaps, before returning to his fireside, he may go and look into a dark cupboard to see if the hyacinths, in fibre, are beginning to sprout. But that represents the sum total of his activity.

I wrote above that, on this first winter, I went mad. For I suddenly said to myself 'I WILL HAVE FLOWERS IN MY GARDEN IN WINTER.' And by flowers I meant real flowers, not merely a few sprays of frozen periwinkle, and an occasional blackened Christmas rose. Everybody to whom I spoke said that this desire was insane, and I suppose 'everybody' ought to have been right. Yet, everybody was wrong. For my dream has come true . . .

I *must* explain my love of winter flowers, in order that the charge of insanity may be refuted.

And yet it is so strong and so persistent – this love – that I sometimes call a halt, and ask myself if it may not be, at least, a little morbid. For there are curious visions that come to me, on blazing summer days, when the garden is packed with blossom like a basket. In an instant, I seem to see the garden bare . . . the crimsons and the purples are wiped out, the sky is drained of its blue, and the trees stand stark and melancholy against a sky that is the colour of ashes. It is then that I see, in some distant corner, the faint, sad glimmer of the winter jasmine . . . like a match that flickers in the dark . . . and at my feet a pale and lonely Christmas rose. And I kneel down quickly, as though I would shelter this brave flower from the keen wind . . . only to realize with a start, that I am kneeling in the sunshine, that there is no flower there, only a few green leaves . . . and overhead, the burning sun.

I wonder why. And yet, perhaps I know. For this passion for winter flowers has its roots deep, deep within me. I have a horror of endings, of farewells, of every sort of death. The inevitable curve of Nature, which rises so gallantly and falls so ignominiously, is to me a loathsome shape. I want the curve to rise perpetually. I want the rocket, which is life, to soar to measureless heights. I shudder at its fall, and gain no consolation that, in falling, it breaks into trembling stars of acid green and liquid gold. I can hear only the thump of the stick in some sordid back yard. The silly thump of a silly stick. The end of life. What does it matter that a moment ago the tent of night was spangled with green and gold? It is gone, now. The colour is but gas . . . a feeble poison, dissipated. Only the stick remains.

I believe that my love for winter flowers has its secret in this neurosis . . . if one may dignify the condition by such a word. I want my garden to *go on*. I cannot bear to think of it as a place that may be tenanted only in the easy months. I will not have it draped with Nature's dust sheets.

That is why I waged this battle for winter flowers. Make no mistake about it. It *is* a battle. There is the clash of drama about it. People think that the gardener is a placid man, who chews a perpetual cud . . . a man whose mind moves slowly, like an expanding leaf, whose spirit is as calm as the earth's breath, whose eyes are as bright as the morning dew. Such ideas are very wide of the mark. A gardener . . . if he is like many gardeners I know . . . is a wild and highly-strung creature, whose mind trembles like the aspen and is warped by sudden frosts and scarred by strange winds. His spirit is as tenuous as the mists that hang, like ghosts, about the winter orchards, and in his eyes one can see the shadows of clouds on black and distant hills.

BEVERLEY NICHOLS, 1932

I must needs add one thing more in favour of our climate, which I heard the king [Charles II] say, and I thought new and right, and truly like a king of England, that loved and esteemed his own country: 'twas in reply to some of the company that were reviling our climate, and extolling those of Italy and Spain, or at least of France: he said, he thought that was the best climate, where he could be abroad in the air with pleasure, or at least without trouble and inconvenience, the most days of the year, and the most hours of the day; and this he thought he could be in England, more than in any country he knew of in Europe. And I believe it is true . . .

There are, besides the temper of our climate, two things particular to us, that contribute much to the beauty and elegance of our gardens, which are the gravel of our walks, and the fineness and almost perpetual greenness of our turf. The first is not known anywhere else, which leaves all their dry walks in other countries, very unpleasant and uneasy. The other cannot be found in France or in Holland as we have it, the soil not admitting that fineness of blade in Holland, nor the sun that greenness in France, during most of the summer; nor indeed is it to be found but in the finest of our soils.

SIR WILLIAM TEMPLE, 1685 (pub 1692)

More and more I am coming to the conclusion that rain is a far more important consideration to gardens than sun, and that one of the lesser advantages that a gardener gains in life is his thorough enjoyment of a rainy day!

MARGARET WATERFIELD, 1907

The whole garden is singing [a] hymn of praise and thankfulness. It is the middle of June; no rain had fallen for nearly a month, and our dry soil had become a hot dust above, a hard cake below. A burning wind from the east that had prevailed for some time, had brought quantities of noisome blight, and had left all vegetation, already parched with drought, a helpless prey to the devouring pest. Bushes of garden Roses had their buds swarming with green-fly, and all green things, their leaves first coated and their pores clogged with viscous stickiness, and then covered with adhering wind-blown dust, were in a pitiable state of dirt and suffocation. But last evening there was a gathering of grey cloud, and this ground of grey was traversed by those fast-travelling wisps of fleecy blackness that are the surest promise of near rain the sky can show. By bedtime rain was falling steadily, and in the night it came down on the roof in a small thunder of steady downpour. It was pleasant to wake from time to time and hear the welcome sound, and to know that the clogged leaves were being washed clean, and that their pores were once more drawing in the breath of life, and that the thirsty roots were drinking their fill. And now, in the morning, how good it is to see the brilliant light of the blessed summer day, always brightest just after rain, and to see how every tree and plant is full of new life and abounding gladness; and to feel one's own thankfulness of heart, and that it is good to live, and all the more good to live in a garden.

GERTRUDE JEKYLL, 1900

Darville, who comes up from the village pub to garden for us, plants potatoes; for it is Easter time and potatoes should always be planted on Good Friday or as near it as possible, be Easter early or be it late.

And then rain falls, a gentle 'growing' rain, as the villagers call it. They look upon it as a friend. To love rain one must live in the country. It falls for several days and the plants strengthen and swell. The faces of the villagers glow as we meet them and discuss it. I listen to it as I lie awake one night. There is at first silence, for the rain has ceased. This silence is so dense as to seem to be black. Then comes the silky rustle of soft rain like the sound of a gentle wind in a ripening corn-field. Some while after comes the steady drip of the rain from the pipes into the rain-water tank. From time to time I hear a far off owl, and once or twice some bird seems to turn in its nest and sleepily grumble. From up the hill comes the raucous sound of a turkey. Soon dawn shows silver grey, and as at a signal several larks rise into the sky, singing in the rain. The grumbles of the sleepy birds cease, and suddenly with the lark the gardenful of birds bursts into song.

CLARE LEIGHTON, 1935

The garden suffers from the long drought in this last week of July, though I water it faithfully. The sun burns so hot that the earth dries again in an hour, after the most thorough drenching I can give it. The patient flowers seem to be standing in hot ashes, with the air full of fire above them. The cool breeze from the sea flutters their droop-ing petals, but does not refresh them in the blazing noon. Outside the garden on the island slopes the baked turf cracks away from the heated ledges of rock, and all the pretty growths of Sorrel and Eye-bright, Grasses and Crowfoot, Potentilla and Lion's-tongue, are crisp and dead. All things begin again to pine and suffer for the healing touch of the rain.

Toward noon on this last day of the month the air darkens, and around the circle of the horizon the latent thunder mutters low. Light puffs of wind eddy round the garden, and whirl aloft the weary Poppy petals high in air, till they wheel like birds about the chimney-tops. Then all is quiet once more. In the rich, hot sky the clouds pile them-selves slowly, superb white heights of thunder-heads warmed with a brassy glow that deepens to rose in their clefts toward the sun. These clouds grow and grow, showing like Alpine summits amid the shadowy heaps of looser vapor; all the great vault of heaven gathers darkness; soon the cloudy heights, melting, are suffused in each other, losing shape and form and color. Then over the coast-line the sky turns a hard gray-green, against which rises with solemn movement and awful deliberation an arch of leaden vapor spanning the heavens from southwest to northeast, livid, threatening, its outer edges shaped like the curved rim of a mushroom, gathering swiftness as it rises, while the water beneath is black as hate, and the thunder rolls peal upon peal, as faster and faster the wild arch moves upward into tremendous heights above our heads. The whole sky is dark with threatening purple. Death and destruction seem ready to emerge from beneath that flying arch of which the livid fringes stream like gray flame as the wind rends its fierce and awful edge. Under it afar on the black level water a single sail gleams chalk-white in the gloom, a sail that even as we look is furled away from our sight, that the frail craft which bears it may ride out the gale under bare poles, or drive before it to some haven of safety. Earth seems to hold her breath before the expected fury. Lightning scores the sky from zenith to horizon, and across from north to south 'a fierce, vindictive scribble of fire' writes its blinding way, and the awesome silence is broken by the cracking thunder that follows every flash. A moment more, and a few drops like bullets strike us; then the torn arch flies over in tattered rags, a monstrous apparition lost in darkness; then the wind tears the black sea into white rage and roars and screams and shouts with triumph, – the floods and the hurricane have it all their own way. Continually the tempest is shot through with the leaping lightning and crashing thunder, like steady cannonading, echoing and re-echoing, roaring through the vast empty spaces of the heavens. In pauses of the tumult a strange light is fitful over sea and rocks, then the tempest begins afresh as if it had taken breath and gained new strength. One's whole heart rises responding to the glory and the beauty of the storm, and is grateful for the delicious refreshment of the rain. Every leaf rejoices in the life-giving drops. Through the dense sparkling rain-curtain the lightning blazes now in crimson and in purple sheets of flame. Oh, but the wind is wild! Spare my treasures, oh, do not slay utterly my beautiful, beloved flowers! The tall stalks bend and strain, the Larkspurs bow. I hold my breath while the danger lasts, thinking only of the wind's power to harm the garden; for the leaping lightning and the crashing thunder I love, but the gale fills me with dread for my flowers defenseless. Still down pour the refreshing floods; everything is drenched: where are the humming-birds? The boats toss madly on the moorings, the sea breaks wildly on the shore, the world is drowned and gone, there is nothing but tempest and tumult and rush and roar of wind and rain.

The long trailing sprays of the Echinocystus vine stretch and strain like pennons flying out in the blast, the Wistaria tosses its feathery plumes over the arch above the door. Alas, for my bank of tall Poppies and blue Cornflowers and yellow Chrysanthemums outside! The Poppies are laid low, never to rise again, but the others will gather themselves together by and by, and the many-colored fires of Nasturtiums will clothe the slope with new beauty presently. The storm is sweeping past, already the rain diminishes, the lightning pales, the thunder retreats till leagues and leagues away we hear it 'moaning and calling out of other lands'. The clouds break away and show in the west glimpses of pure, melting blue, the sun bursts forth, paints a rainbow in the east upon the flying fragments of the storm, and pours a flood of glory over the drowned earth; the pelted flowers take heart and breathe again, every leaf shines, dripping with moisture; the grassy slopes laugh in sweet color; the sea calms itself to vast tranquillity and answers back the touch of the sun with a million glittering smiles.

CELIA THAXTER, 1894

'You're going to *Tulsa?*' the woman on the plane said in a tone of utter disbelief. 'On a *Sunday?*' And she was going to Dallas. Tulsa is like Basingstoke – not a place anyone takes seriously unless they happen to live there. Work had brought me unavoidably to this city, in fact a soulless conurbation, spread like margarine across the flat, dry cracker of Oklahoma. Contrary to popular belief, however, there *was* something to do on a fine Sunday in early May, since the Tulsa Area Iris Society was holding its Spring Flower Show at The Garden Center in Woodward Park, reached (unless you drive, as everyone in Tulsa does) after a dispiriting three-mile walk along the much-hymned Route 66, past gas stations, used car lots, drive-thru' fast-food outlets and The Flower Factory, which offers 'Hospital Bouquets' and 'Sympathy Pieces' at a discount ('Say Everything with Flowers').

Real gardeners in Tulsa are brave people. The soil is a heavy clay, which gets waterlogged in the spring rains and bakes absolutely solid during the long hot summer. It is possible to keep a garden going through the summer months, I was assured, but it requires gallon upon gallon of (metred) water. My informant's husband was an aeronautics electrician and had rigged up a self-watering system. Most people don't bother. Azaleas do well, as do roses, which bloom early, and may be seen in all their garish splendour in the famous rose garden in Woodward Park. Honeysuckle also seems to thrive, but other plants have to take their chance. There is, however, one splendid exception. 'Iris are one of the easiest perennials to grow,' advises a promotional sheet put out by the Iris Society. 'They survive with less care and reward you with fine blooms with a minimum of attention.' Not that the Society's members leave their plants to fend for themselves. Fred Smith, a veteran of the Tulsa Iris Growers' Association, confided that, having cultivated and cut one's specimen, 'grooming' was quite in order, indeed expected. 'They can do just about anything,' he said – although someone who attempted to improve a bloom with water-colours was rather frowned upon. The Iris Show takes place when the tall bearded variety is at its peak – though real fanatics like Mr Smith will travel through several states in order to prolong the season's view-ing: he had just been in Texas, where the blooms would now be fading, and was intending to go on up to Oregon later in the month, to catch the crop of the northern growers.

The show's organizers provide uniform receptacles rather like specimen bottles, so that judges will not be distracted by inferior blooms shown off to advantage in Great Aunt Jemima's best lead crystal. In each container a single stem is placed, held erect by assorted wedges. There is a prize for each variety, so that most growers are likely to collect at least one award: 'But the first prize is the only one that means anything,' I was told firmly. Mistakes are occasionally made, and this year a renowned grower had entered, and been given a prize for, a specimen labelled 'Best Bet'. 'Too light,' said Mr Smith, scribbling 'Misnamed' on the label. '*She* should have known, and the judges surely should have.'

Once you have chosen your exhibit and correctly identified it – to an untutored eye some varieties seem to be distinguished from one another by the minutest details in beard or fall, and even the judges have a directory to consult in disputed cases – you can start titivating it, rather in the manner of contestants in a beauty pageant. If there is blight or pest damage on the foliage, you may 'remodel' the edge of the leaf with a scalpel, so long as you do not cut away too much and end up with one entirely the wrong shape. Calyces which have faded to an unsightly papery brown are meticulously trimmed, and much leaf-polishing takes place. Points are awarded for balance and number of flowers unfurled and in bud; the shape of the stem (Mr Smith demon-strated the sort of reflex curves which attract high scores); and the way the stem branches. A week or so before the show, many exhibitors insert polystyrene wedges between bud and stem of the growing plant to encourage a graceful shape, removing these on the day. Another trick is to cut your stem just as it reaches perfection and pop it into the refrigerator. With luck, it will remain in top condition for up to a week, before being rushed to the judges' benches. Such cryogenic techniques have their drawbacks, however, and your timing has to be precise. Preserved specimens have made it to 'The Queen's Table', only to collapse, rather in the manner of a lettuce, and be unceremoniously returned to take their place among *hoi polloi*.

Curls, lace frills and branching have improved dramatically over the last five years, said Mr Smith, pointing to a white and lilac confection called 'Fancy Tales', with khaki falls and a burnt-orange beard. 'Ain't nobody had one like *that* fifty years ago,' he assured me. 'It's a dog, mind you. Blooms once in every five years – if you can keep it alive that long. Heck, you oughta get a blue ribbon just for growing it and bringing it along.' The judges appeared to agree, but 'Fancy Tales' did not make it to The Queen's Table, not even as a hand-maiden. Neither did 'Isn't This Something', which certainly was: a repellent object with marbled purple-cream petals and a virulent ginger beard. Most of the 'Finalists to the Queen's Court' were in shades of lilac and white, apart from 'Edith Wolford', a distinguished iris with chrome yellow petals, mauve falls and a yellow beard. 'Queen of the Show', deservedly, was a fine specimen of 'Clear Fire', an extremely handsome flower in assorted shades of chocolate, from buds the colour of darkest Bournville to falls of dusty cocoa.

PETER PARKER, 1994

Whether awake or asleep, my thoughts and dreams are dominated by the prospect of Chelsea. Yet now the day-to-day routine of running the nursery and garden smoothly must take precedence.

When you see the spectacular displays at this famous show, you might think that exhibitors have nothing else to do but plot and plan to make them so. Far from it. Although it may be chosen as the loveliest time of the year [the third week of May], the Chelsea Flower Show could not come at a more inconvenient time. Everything – just everything – inside and out, is screaming for attention at this most hectic time in the plantsman's calendar . . .

I do not enjoy the waiting days before Chelsea. Most of the plants look as I would wish them to be, but a few are still reluctant, even though I have moved them back and forth from the warmer plastic house to the cool shade house as the weather has changed. A heat-wave would be a disaster, but just a few degrees more of warmth would be gratefully acknowledged by plants and people . . .

It is the weekend before Chelsea. I am longing to start selecting the best plants and to pack them into boxes and trays ready for the journey to London. Now that everything is so nearly ready I find it hard to concentrate on other matters. The two long tunnels of plants look to me like a scattered jigsaw puzzle that is waiting to be put together to make a picture I could not possibly put down on paper, but which will form, I hope, in my head.

At last it is Monday morning, 13 May, just four days before we take our first load of plants to Chelsea. It is raining but not cold. Suddenly it is growing weather and the birds sing without stopping. An extra-large trestle table has been brought into the office for Lesley and Winnie Dearsley who are preparing 5,000 catalogues. Each must have a separate price list and order form put inside. Then they are made into bundles of fifty, the bundles packed into plastic fertilizer sacks (obtained from our neighbouring farm) because they will (we hope) be stored underneath a rose exhibitor's stand, near to us. When the roses are watered we cannot risk water running into cardboard boxes. Rosie is checking our tickets and passes to enable us to get into Chelsea, while Georgina Cherry is taking care of the daily postal routines.

At such times I remember with amusement my brief career as a school-teacher, when the classroom buzzed like a beehive with everyone engaged in a variety of projects.

Our trolleys and watering cans are painted clearly with 'CHATTO'. We are pleased to let people borrow them but we like to see them returned. Thin, split bamboo canes are used to hold the plant names and these are being painted matt-black. They will scarcely show among the foliage.

At last, Sue and I could start to box the plants. We removed dead or damaged leaves, carefully tied those which needed support, and kept plants which needed similar conditions in the same boxes. (It is exasperating when I am staging to have both dry and damp-loving plants in the same box.) Most of the plants look well, with healthy fresh foliage, and many are in flower, some in bud or just opening. In a week's time all the anxiety and pleasure of doing it will be over. On the practical side, Keith is checking last year's list of all the odds and ends: the materials needed to make our little pool, large sheets of plastic to cover the site, rakes to clear the grass around our stand when we have finished, a dustpan and brush, small brushes to dust off the leaves, scissors, string, many pieces of dark material to hide our large containers, and so on. Large plastic fertilizer bags are filled with crushed bark to finally dress the stand. If a new idea to improve our methods comes up during the show we write it down and put it into our Showbox to remind us. Once Chelsea is over we turn to a thousand other problems which have been waiting for our attention.

It took two of us almost three days to select and prepare our plants. The beautiful large rosettes of *Verbascum* and *Onopordum* were carefully folded and tied upright so that the perfect shape and texture of the leaves would not be damaged. Other growers will be wrapping delicate heads of flowers such as *Iris*, *Delphinium* and orchids in cotton wool. Many plants have been rejected, but there still seems to be a great number to be packed into the vans.

The first day of staging arrives at last. It is a lovely warm, sunny morning. Two large hired covered vans arrive early, driven by the owner and his son who have always transported our plants to Chelsea with great care. Yet still I am anxious until I see the great pots of ornamental rhubarb, *Rheum palmatum*, unpacked and standing in London as perfect as when they were in my tunnels. Many hands are there to lift the heavy trays and boxes. I hover around making sure that all the right plants are packed. Then, Keith and I leave in our own van filled with all the practical accoutrements. As always, the journey is slow because of the volume of traffic into London. We pass one of our convoy halted in a lay-by with a puncture. 'They can manage,' says Keith. 'We will go on and prepare the site.' This was good judgement, since they did manage and we had laid the sheets of plastic sheeting to protect the grass site and emptied our van before the plants came in. As the men brought in the trolleys loaded with plants I directed where they should be put so that the drought-loving plants were stood along the 'dry' side of my stand and the damp and shade-loving plants in their appropriate positions. There seemed an awful lot of plants already, but the site, flat and empty, 3oft × 2oft (10m × 6m) looked very large. It would easily be dwarfed by a few large rhododendrons, but it takes many herbaceous plants to create a living garden-scene. Once we are on our way I always enjoy the first morning's drive to Chelsea. The countryside looks fresh and leafy. This year the hedges are not so loaded with creamy masses of hawthorn blossom as on previous occasions but apple and lilac trees are in bloom, with laburnum still to come. As we cross Chelsea Bridge I see the familiar blue and white striped canvas of some of the marquees and my heart gives sudden jumps of apprehension and anticipation. So far, despite the winter's worries, our plants look satisfactory, but the last phase of this adventure, the staging is still, for me, the unknown element. The principles behind my planning remain the same, but each year the plants behave differently and I, too, have changed, so the combinations are never the same.

But always the same is the pleasure we have as we enter the great marquee and find some of our old friends already there. The feeling between plantspeople is genuine. You may not see someone for a year

– or several years – but the feeling of kindred spirit remains. It is very heartwarming. On my return home that evening there are last-minute things to be checked on the nursery, in the office and finally in the house. Clothes and food sufficient for three days' staging are packed.

Next morning I am in the garden early to find plants for my little pond and to find a few extra-large hostas for another exhibitor who had worries yesterday.

By midday, with two more vanloads of plants, we were back in London. With everything unpacked there was scarcely room to move around the stand because of the bewildering sea of plants; too many for comfort I feel sure, and yet I can see at once that we lack really bulky material for the central groups. The winter losses cannot be made up in one season.

I do not really enjoy the first day of staging, trying to find the main thread of my ideas which will lead me through the next three days; but first I stage the trees and shrubs which are needed to give height to the two main groups. Where and how they will be stood is vital because they set the proportion for the rest of the design. Almost invariably, most of them need to be stood on boxes, or on buckets, to achieve the overall height and bulk needed in relation to the length of the stand.

The 'pond' is put in place below a group of willow and bamboo which are lightened with the delicate pink, white and green tinted young leaves of *Acer negundo* 'Variegatum', and with spreading branches of gold and white forms of variegated dogwood. As each pot is placed it must be watered well and the name attached so it will be seen from a distance and the pot and boxes concealed. By evening I was less than happy, but knowing from experience that a night's rest often produces the solution next morning I was glad to be driven away to our hotel by David Ward who had been my calm and helpful assistant all day. Next morning he resited the pond, giving us more space to build up a design of ligularias, Solomon's Seal and the beautiful white *Dicentra spectabilis* 'Alba', which brought the white tints of the variegated dogwood down to the edge of the back of the little pool. Yet still there was a link missing because the dramatic green and white foliage of a fine Japanese grass *Miscanthus sinensis* 'Variegatus' was not, this year, sufficiently bold to make the feature I needed. Suddenly I picked up a very tall clump of *Arundo donax* 'Variegata' whose long ribbon-like leaves cascaded from strong stems over six feet tall. I had intended to use it properly among my Mediterranean plants because it comes from the warm south and can only be used in British gardens if bedded out in summer. Why not use it as we might use a *Canna* or *Phormium* bedded out in summer near the pool? My conscience salved, I put it into place and immediately knew that it was the piece I had been looking for. The rest of the day was spent with hostas, primulas, water iris and other moisture loving plants, fitting together interlocking groups of colour and texture to complement each other.

Around us the marquee was filling as more and more exhibitors were arriving and quietly unpacking; everyone was totally absorbed in the final stages of this most important gardening event of the year. Whether hauling fully-grown trees and shrubs or exotic orchids, there was the same degree of caring, concern, consultation and relief as the teams worked together and the designs took shape; everyone doing their damnedest out of love for their plants.

By the end of the day, about 7 p.m., we were glad to relax in hot baths, change our dirty clothes and have a good meal. Before bed I telephoned Andrew, to give him love and good wishes for his birthday today.

It is Sunday morning and we are among the first to arrive, to find blackbirds and thrushes pulling fibrous strips from my pieces of bark and practically queuing up to help themselves from a sack of moss. All day they snatch pieces of wood fibre, sometimes from beneath my hands, then fly off to make nests in other exhibitors' shrubs. By the end of the show, eggs are laid.

Rosie and Madge have arrived to help with labelling and watering, and provide a hundred and one little services throughout the day which ease the strain of staging.

The Mediterranean side of my stand has worried me most. While I have a few fine large shrubs remaining from the winter's losses, I still have to create the effect of established plants by grouping several together and lifting them to appropriate heights. I can scarcely eat, or sit down, fearing to be distracted from putting together the groups which are in my head and which I have yet to find among the sea of boxes. Then, suddenly, the current seems to be switched off; the design looks penny-plain. I have to walk away, to rest my mind and eye, just long enough to refocus.

By 7.30 on Sunday evening almost the last piece of the puzzle had been found. But not quite: two pieces were missing, and suddenly I could not see them among all the bits and pieces remaining, nor had I the energy to kneel and get up once more.

Back in my room it was bliss to peel off soaking wet trousers and shoes, to lie as though weighted with lead in a long hot bath, and afterwards to go out and eat together with my young staff who have shared all the emotional ups and downs of creating our exhibit. But that night in bed my feet and legs throbbed from kneeling day after day, and my left hand was sore and swollen from placing and planting.

On Monday all exhibits must be finished, completed before 4 p.m. and all rubbish removed. As I walked towards my stand early after breakfast two plants suddenly appeared before me, tall blue polemoniums. Jacob's Ladder had opened overnight, just what I needed with purple-leafed *Viola labradorica* to make a carpet beneath them. Finally we finished planting either side of the wood-block path of circular wooden 'pennies' which winds through the centre of our exhibit, with ferns and hostas, completely screening the two wooden 'caves' hidden in the centre of the stand where we can store extra catalogues and our personal belongings. Some exhibitors have room for a summerhouse for these necessities. For us, it is a problem we must solve in a different way. Primarily, we depend on the generosity of a neighbouring rose stand whose bowls are lifted on tabling, leaving hiding-places between canvas-covered trestles beneath. By 11 a.m. our large hired van had arrived with Keith and Kazu ready to clear away the debris. Anxious to the last, I am found brushing away the last crumbs of crushed bark clinging to any leaves which, if left, would spoil the fresh untouched look I have been striving these past months to achieve. At the same time I am trying to talk politely to an endless stream of visitors, mostly journalists, photographers, radio interviewers, and occasionally visitors from abroad. In the mêlée David is putting the final touch, tacking long pieces of rugged bark to hide the edging board which surrounds the stand. This very simple finish seems just right, partly concealed by foliage creeping over its edges. Last of all, a light mist of water applied with the thumb held over the end of the hosepipe removes the slightest film of dust and restores a true country garden freshness. Overall, the colours of flowers and leaves are soft and gentle, accented or shadowed here and there with darker tones. Thankfully, I can leave the clearing up and return to my room for a short rest. I lie on my bed, adrenalin still throbbing through my veins, yet there is also a deep feeling of relief. Whatever the outcome may be, we have managed to stage another Chelsea and I feel relatively content, to emerge in the late afternoon as clean as bath and nailbrush can make me.

It is a great moment at last, to be dressed for the occasion; to be calmed by a walk through the beautiful buildings and courtyards of the Chelsea Hospital and then into the swept and sprinkled show ground. Not a dead leaf or speck of rubbish remains; every exhibit is complete, with the traditional white ropes set in position, in an attempt to prevent the public falling over the exhibits and exhibitors in their enthusiasm to see everything. Naturally, I go straight to our exhibit to see it properly for the first time, without the piles of empty boxes, damaged or unwanted plants, and all the rest of the paraphernalia which collects around us.

I am always impressed by the layout of exhibits in the main marquee which covers nearly four acres and is, I think, the biggest in Europe. Usually the Chelsea schedule, rules, regulations and application form drop through the letter box before Christmas and we are asked to send the measurements of the exhibit we would like to make. Then, sometime in January, in the depths of winter, the plan of all the allocated exhibition sites appears and already Chelsea fever is incubating. But much care is taken to see that exhibits are sensitively grouped. We do not always have the same neighbours. This is good, because we meet new people, and by the end of the show we have made new friends.

Many people imagine that there must be great rivalry, both for awards and for customers, but this is not so. There are several judging committees consisting of about twelve people each, all experienced and knowledgeable in horticulture. Dark-suited and inscrutable they enter the marquee with papers and pens poised. There are no 1st, 2nd and 3rd class awards as we see them in country shows. If two exhibitors in the same section produce a stand which the judges consider to be worthy of a gold medal then two medals will be awarded. This means that we are not competing against each other, but strive to do as well or better than our previous exhibit. The standard of excellence can be, and is, raised each year. This may not be so in the case of every exhibitor but for many it is, and this makes the Chelsea show a perpetual challenge for us all.

I wander around the widely varied exhibits with time and space to admire them and talk to old friends, also transformed for this special evening. Suddenly there are little flurries of movement; photographers dash around like children playing hide and seek between the banks of flowers and you realize that the royal parties have begun to arrive. Conducted by important members of the Royal Horticultural Society Committee, several little groups make stately progress, often hidden from view until they suddenly appear beside you, while blackbirds and thrushes continue to trill never ceasing fanfares ahead of them. Hearts beat when we are presented to each of the royal parties, but above all to the Queen. Hands are touched, I make a curtsey (I hope I remembered), some words are exchanged, and the moment has passed. I am left light-headed by the great sense of occasion, but when I feel the ground beneath my feet once more I am conscious of admiration and concern for our Queen, who finds something kind and encouraging to say almost every day of her life to people she meets, concerning their many and varied activities.

Finally, the last party leaves and we may go home. I suddenly feel cold and very hungry. Finding my way back to where we are all staying, I hurry into the dark warmth of our favourite little eating place. My friends have eaten, but are waiting expectantly to ask: 'Who did you meet, tell us about it.' We all share the fun and excitement, while I

savour the relaxed atmosphere and a comforting meal. For the first
time in a week, I fall into bed ready to sleep, exhaustion blotting out
the ceaseless hum of the north-bound traffic nearby.

With the pressure off, I slept heavily, dreaming wild dreams till 6.30
a.m., when a tap at my door had me down to breakfast by 7 a.m. David
drove us to the main entrance to let the girls hurry to our stand, while
we had to drive round the hospital grounds to the car park. Already, at
8 a.m., the pavements are three deep with crowds, more people hurrying from the Underground station to join the queue on this first day –
Members' Day – which is perhaps more crowded than any other
during the week. Our usual route to the car park was cut off by road
works so it took more time than usual to park and find our way, again,
round the stately hospital building and the already crowded avenues to
reach our stand by 8.30 a.m. We arrived to find a wall of people around
it, hastened to pick up our order books and catalogues and turned to
face our audience. Suddenly as though the light had gone out, it
became so dark we could hardly see the colour of the flowers. Lightning and thunder followed almost simultaneously and torrential rain
hammered on the billowing canvas above our heads. (I suddenly remembered a story told me by Joe Elliott, about his father, Clarence
Elliott, who never entered the Chelsea Flower Show tent without a
small, very sharp knife. I also thought of all the luckless exhibitors
outside who have been rained on throughout the preparations and
must suffer yet more soaking.) Visitors suddenly crowded into the big
marquee and the congestion was almost unbearable. But soon the rain
stopped, followed by sunshine for the rest of the day.

We all need the enthusiasm expressed for our stand; it affects us like
champagne when we have been talking, taking orders, giving advice,
selling catalogues, answering the occasional silly question and greeting
friends from past years for hours, it seems, on end. Just when we think
that we are totally drained, a particularly sympathetic visitor not only
gives us a great lift, but kindles enthusiasm; sometimes even coerces
those around him (usually her) to buy a catalogue, or to visit the
garden. We all have a great laugh and everyone feels revived.

The day passed quickly in this euphoric atmosphere – and we did eventually discover that we had been awarded a Gold Medal, our tenth in eleven years; such a feeling of relief now, but not of excitement; only the public can produce that: the pressing crowds stationary around us, the comments which convince us that what we have done is giving pleasure.

BETH CHATTO, 1988

We hear much of green fingers, but nothing of green brains. Clearly, however, it is possible sometimes to learn things that one is not born with. That this is true of gardening is well illustrated by a friend of mine who had never gardened in her entire life until suddenly inspired by a large new greenhouse her husband had bought himself. When she invited me to view the first results of her labours I was astonished to find an almost professional array of flowers, seedlings, boxes of annuals, tomatoes and so on. When I expressed astonishment, to which I added much congratulation, she made the sensible and disarming reply: 'Well, I can read, can't I?'

H. E. BATES, 1971

1. If you prune your vines the moon in full, and posited to Taurus, Leo, Scorpio, or Sagittary, neither worms nor buds will infest your grapes.
2. Trees are not to be grafted the moon waning, or not to be seen.
3. Cut what trees you would have quickly grow again, when the moon is above the earth, in the first quarter; and if it may be, joyned to Jupiter or Venus.
4. Sow or plant when the moon is in Taurus, Virgo, or Scorpio, and in good aspect of Saturn.
5. Set or sow all kinds of pulse the moon in Cancer.
6. Dress your gardens, and trim your small trees and shrubs when the moon is in Libra or Capricorn.
7. Set or cut any tree or shrub, that you would have its growth retarded, in the decrease of the moon in Cancer.
8. Set, cut or sow what you would have speedily shoot out again, or spring and grow, in the increase of the moon.
9. When you sow to have double flowers, do it in the full of the moon; and when the plant is grown to a bigness fit to be removed; do it also in a full moon, and as oft as you transplant them.
10. Neither graft, set, sow or plant any thing that day whereon there happeneth an eclipse either of sun or moon, or when the moon is affected by either of the infortunes Saturn or Mars. I might have given you many more; but these may be sufficient at present.

SAMUEL GILBERT, 1682

CHAPTER FOUR

Plants in the Garden

Auriculas in the Greenhouse
Bryan's = Ground =
Simon Dorrell

On the 13th of November 1647 the King [Charles I] crossed the sea, was safe landed at Cowes in the Isle of Wight, where Colonel Hammond, the Governor was attending, and passed through Newport (the principal town in that Island). The Governor, with alacrity and confidence, conducted his Majesty to Carisbrook Castle, attended only by Sir John Berkeley, and two gentlemen, his servants. Sure I am, many that cordially loved the King, did very much dislike his going to this place, it being so remote, and designed neither for his Honour nor safety; as the consequence proved. A gentlewoman, as his Majesty passed through Newport, presented him with a damask rose which grew in her garden at that cold season of the year, and prayed for him, which his Majesty heartily thanked her for.

SIR THOMAS HERBERT, 1678

PLANTING FLOWERS ON THE EASTERN
EMBANKMENT

(Written when Governor of Chung-chou)

I took money and brought flowering trees
And planted them out on the bank to the east of the Keep.

I simply bought whatever had most blooms,
Not caring whether peach, apricot, or plum.
A hundred fruits, all mixed up together;
A thousand branches, flowering in due rotation.
Each has its season coming early or late;
But to all alike the fertile soil is kind.
The red flowers hang like a heavy mist;
The white flowers gleam like a fall of snow.
The wandering bees cannot bear to leave them;
The sweet birds also come there to roost.
In front there flows an ever-running stream;
Beneath there is built a little flat terrace.
Sometimes I sweep the flagstones of the terrace;
Sometimes, in the wind, I raise my cup and drink.
The flower-branches screen my head from the sun;
The flower-buds fall down into my lap.
Alone drinking, alone singing my songs
I do not notice that the moon is level with the steps.
The people of Pa do not care for flowers;
All the spring no one has come to look.
But their Governor General, alone with his cup of wine,
Sits till evening and will not move from the place!

PO CHU-I, 819 (trans. ARTHUR WALEY, 1946)

The Names of several Herbs, &c. fit to set Knots with, or to edge Borders to keep them in fashion, &c.

Dutch or French Box, it is the handsomest, the most durable, and cheapest to keep.

Hyssop is handsom, if cut once in a fortnight or three weeks in the growing season.

Germander was much used many years ago, it must have good keeping.

Thrift is well lik'd of by some, it is apt to gape and be unhandsom.

Some use Gilden-Marjoram, or Pot-Marjoram with good keeping will be handsom.

Also besides the fore-named, you may edge Borders with divers things; as Pinks, they will be very handsom by cutting twice a year.

Violets double or single, they will thicken and be handsom if oft cut.

Grass cut oft.

Periwinkle cut oft.

Some use Lavender-Cotton, and Herba-grace, &c. will be handsom if kept well.

Rosemary may be kept low as other herbs, if oft cut.

Lavender as it may be kept, will be both low and handsom.

Sage likewise.

Primroses and Double-Daisies are set for that purpose likewise, but they ought to be planted something shady.

Another thing I thought good to mention: It is common in the mouths of many, that Box doth take away all the heart of a ground where it grows; but the naked Truth is, that it doth not draw so much vertue from a ground as other herbs doth; my reason is, because it doth not grow so fast, and so by consequence not draw so much vertue from the place where it grows; and in case it do begger or barren a place where it grows, it comes to pass by its long standing compleat and handsom, which is a part of its excellency; it being the most durable of any kind of herb wherewith Knots are made; but to prevent, or rather amend the inconveniences that seem to follow by the running of its roots into your Knot, which any other herb doth much more, by how much other herbs do grow more than Box; the remedy is, with a knife or piece of an old Sithe once in two years to cut the root down close to the Box on the inside of your knot, and then if need be to new flourish your work with little fresh mould; also take notice that often cutting of either Box or any other herb, is a means to prevent the much running of the roots into your knot, the same reason is for hedges or borders.

LEONARD MEAGER, 1688

EUTOPIA

There is a garden where lilies
    And roses are side by side;
And all day between them in silence
    The silken butterflies glide.

I may not enter the garden,
    Though I know the road thereto;
And morn by morn to the gateway
    I see the children go.

They bring back light on their faces;
    But they cannot bring back to me
What the lilies say to the roses,
    Or the songs of the butterflies be.

FRANCIS TURNER
PALGRAVE, 1871

Our summer in Maryland, (1830), was delightful. The thermometer stood at 94, but the heat was by no means so oppressive as what we had felt in the West. In no part of North America are the natural productions of the soil more various, or more beautiful. Strawberries of the richest flavour sprung beneath our feet; and when these past away, every grove, every lane, every field looked like a cherry orchard, offering an inexhaustible profusion of fruit to all who would take the trouble to gather it. Then followed the peaches; every hedge-row was planted with them, and though the fruit did not equal in size or flavour those ripened on our garden walls, we often found them good enough to afford a delicious refreshment on our long rambles. But it was the flowers, and the flowering shrubs that, beyond all else, rendered this region the most beautiful I had ever seen, (the Alleghany always excepted.) No description can give an idea of the variety, the profusion, the luxuriance of them. If I talk of wild roses, the English reader will fancy I mean the pale ephemeral blossoms of our bramble hedges; but the wild roses of Maryland and Virginia might be the choicest favourites of the flower garden. They are very rarely double, but the brilliant eye atones for this. They are of all shades, from the deepest crimson to the tenderest pink. The scent is rich and delicate; in size they exceed any single roses I ever saw, often measuring above four inches in diameter. The leaf greatly resembles that of the china rose; it is large, dark, firm, and brilliant. The sweet brier grows wild, and blossoms abundantly; both leaves and flowers are considerably larger than with us. The acacia, or as it is there called, the locust, blooms with great richness and profusion; I have gathered a branch less than a foot long, and counted twelve full bunches of flowers on it. The scent is equal to the orange flower. The dogwood is another of the splendid white blossoms that adorn the woods. Its lateral branches are flat, like a fan, and dotted all over with star-like blossoms, as large as those of the gum-cistus. Another pretty shrub, of smaller size, is the poison alder. It is well that its noxious qualities are very generally known, for it is most tempting to the eye by its delicate fringe-like bunches of white flowers. Even the touch of this shrub is poisonous, and produces violent swelling. The arbor judae is abundant in every wood, and its bright and delicate pink is the earliest harbinger of the American spring. Azalias, white, yellow, and pink; kalmias of every variety, the too sweet magnolia, and the stately rhododendron, all grow in wild abundance there. The plant known in England as the Virginian creeper, is often seen climbing to the top of the highest forest trees, and bearing a large trumpet-shaped blossom of a rich scarlet. The sassafras is a beautiful shrub, and I cannot imagine why it has not been naturalized in England, for it has every appearance of being extremely hardy. The leaves grow in tufts, and every tuft contains leaves of five or six different forms. The fruit is singularly beautiful; it resembles in form a small acorn, and is jet black; the cup and stem looking as if they were made of red coral.

FRANCES TROLLOPE, 1832

My learned and travelled friends who tell me I cannot naturalize Narcissus in thick grass, will hardly say we cannot grow our own lovely British tree willows, or have our own native Heaths in all their delightful variety growing near us in picturesque tangles, and some of our own more beautiful Wild Roses in the hedge! The passion for the exotic is so universal that our own finest plants are never planted, while money is thrown away like chaff for worthless exotic trees like the Wellingtonia, on which tree alone fortunes have been wasted. Once on the bank of a beautiful river in Ireland, the Barrow, I was shown a collection of ornamental Willows, and very interesting they were, but among them not one of our native Willows, which are not merely as good as any of the garden Willows but as good in beauty as the Olive tree – even where the Olive is most beautiful. We search the world over for flowering shrubs – not one of which is prettier than the Water Elder (Viburnum Opulus), common in Sussex woods, and often seen near the water-side in Surrey. Mr Anthony Waterer, who has the finest nursery in England in our own day, told me that when asked for a number of it he could not find them in his own nursery, or in any other . . . However well people may know the beauty of our fields and woods in spring or summer, few have any idea of the great number of flowers that are wild in our own country, and worth a home in gardens – at least in those of a picturesque nature. Few of us have much notion of the great variety of beauty that may be culled from British flowers alone. Many of us have full opportunity of seeing the beauties of the fields and hedges; not so many the mountain plants, and few, such rare gems as Gentiana verna, which grows wild in Teesdale, and here and there on the western shores of Ireland; or the mountain Forget-me-not, a precious little dwarf alpine that is found but rarely in the north. It is only by a good choice of the plants of the British Isles that we can hope to arrive at a 'garden of British plants'.

It is not only the curious and rare that may afford us interest among the plants of Britain; among them are plants of much beauty. Even for the sake of plants for lakes, rivers, ponds in parks, pleasure grounds, or gardens, the subject is worthy attention. For the rock-garden, too, many of our wild flowers are fitted. In any part of the country where the soil or surface of the ground suits the habits of a variety of native plants, it would prove interesting to collect kinds not found in the neighbourhood, and naturalize them therein; and wherever the natural rock crops up, much beauty may be added by planting these rocky spots with wild mountain flowers.

WILLIAM ROBINSON, 1870

I remember so well the occasion on which I first made contact with the night-scented stock of Libya. We were coming in from a long and wearisome patrol in the heart of the desert during the Senussi campaign of 1916, and I, in common with every newly-arrived soldier who is called upon to serve in Libya, then entertained a hatred of its gross sterility, the harsh outline of its limestone crags, and its thirteen hours of unbroken, burning sun every day. It was not until much later that I became a desert addict and could see its beauty at dawn and sunset, when the purple shadows on the high scarps show up striking rose-coloured features in the flat blur, and could appreciate the invigorating quality of its pure air. Homesick for England and its moist greenness after a burning day, and above all for the summer scents of English gardens, I drove on northwards towards the coastline, where the harsh scrub of the desert yields up grudgingly to the richer growths of the rainfall belt. Then, as I topped a ridge in the dusk, I imagined I had jumped some 2,000 miles northwards, for a wave of unmistakable night-scented stock perfume suddenly impregnated the air – and the scent of the small desert variety is far more powerful and, to my mind, sweeter than that of our English variety.

I stopped the car, the remaining vehicles of the patrol pulling up behind me, and from the darkness came a voice with the Devon accent: 'Gawd, smell that – we're back in old England!' Then, in the light of the headlamps I saw by the side of the track a mass of the pale mauve flowers of the night-scented stock, so inconspicuous and unassuming by day, and so shining and starlike at dusk, and I realized that Libya, despite its harshness under the sun, had its softer moments after nightfall, and that a desert which can produce night-scented stock is not entirely hopeless.

MAJOR C. S. JARVIS, 1951

One of my happiest botanical memories is of a seemingly un-distinguished olive grove on Rhodes in the last days of October. In melancholic mood (for I had made that great mistake of being preoccupied with home thoughts abroad), I had walked several miles along dusty lanes through a harsh, dry autumn landscape, the air rich with the overpowering smell of pressed olives. Suddenly my spirits lifted at the sight of the bare soil beneath the olive trees pale blue with the spikes of tiny *Muscari parviflorum*, the only autumn-flowering grape hyacinth. Golden autumn sunlight streamed past gnarled trunks and through silvery-grey leaves and all was still. Later, as I rested my legs and feet and drank welcome beer beneath a yellowing vine in the garden of a roadside taverna, to await the afternoon's ancient bus back to Rodhos town, and as the taverna-owner and his family sorted through their last gathering of grapes, I felt one of those curious surges of contentment that life provides from time to time. Not least, I had been fortunate to see a plant with which few botanists or plantsmen are familiar, and in profusion; but also I had been in the right place at just the right time, and on a warm and sunny day, to witness one of nature's less flamboyant but most perfect displays.

JOHN AKEROYD, 1991

Here [at Ning-po], as at other places, I made many inquiries after the supposed yellow camellia, and offered ten dollars to any Chinaman who would bring me one. Anything can be had in China for dollars! and it was not long before two plants were brought to me, one of which was said to be light yellow, and the other as deep a colour as the double yellow rose. Both had flower-buds upon them, but neither were in bloom. I felt quite certain that the Chinaman was deceiving me, and it seemed foolish to pay such a sum for plants which I should in all probability have to throw away afterwards; and yet I could not make up my mind to lose the chance, slight as it was, of possessing the *yellow camellia*. And the rogue did his business so well. He had a written label stuck in each pot, and *apparently* the writing and labels had been there for some years. I fancied I was as cunning as he was, and requested him to leave the plants and return on the following morning, when he should have an answer. In the meantime I asked a respectable Chinese merchant to read the writing upon the labels. All was correct; the writing agreed with what the man had told me; namely, that one of the plants produced light yellow blooms, and the other deep yellow. 'Did you ever see a camellia with yellow flowers?' I inquired of my friend the merchant. 'No,' said he, in his broken English. '*My never have seen he, my thinkie no have got.*' On the following morning the owner of the plant presented himself, and asked me if I had made up my mind upon the subject. I told him that I would take the plants to Hong-kong, where I was going at the time; that they would soon flower there; and that, if they proved *yellow*, he should have his money. This, however, he would not consent to; and at last we compromised the matter, I agreeing to pay half the money down, and the other half when the plants flowered, providing they were 'true'. On these conditions I got the camellias, and took them with me to Hong-kong. It is almost needless to say that when they flowered there was nothing yellow about them but the stamens, for they were both semi-double worthless kinds.

ROBERT FORTUNE, 1847

The rambling days went by in expeditions hither and thither up among the breasts of Axeinos and Thanatos, and over the enormous alps, in search of seed: to end always in an ecstasy of happiness, as we sat at rest in the tent door, absorbing the panorama of the fells, browning to autumn, while away over the Clear Lake, beyond its mountain barrier, the distances opened out infinitely far, over the crumpled lower ranges away below, in ripple over world-wide ripple of softest and most entrancing blue, with the gaunt nakedness of the Gadjur range for crest of its culminating wave. Often Gadjur bred level banks of violet darkness in the afternoons, and Thanatos, not to be outdone, ere long filmed over the glory of the day in dullness, and out of dullness ere long brewed dark, which became violent, became a torrent of rain, became deluges of snow that fell in soft cascades all night. I would wake from fitful slumbers to what sounded like alarming stone-falls from Croda Rossa or Axeinos, possibly threatening the camp; but on looking out it was only Bill and the servants clearing the solid masses of snow off the tents. And so would dawn, in time, a new day of stark diamond-clear radiance, almost unbearably brilliant on a new world of whiteness, which, however, rapidly began to resume its normal colours under the hot lances of sunlight.

Off one would start, this way or that; and on return would find the camp in a mighty fluster of consternation, black-and-violet layers of storm having by now accumulated over the whole northerly region, soon to materialize over us also, in furious squalls, thunder and lashing hail, together with the penetrating grisly cold that floods all these high alps the moment that the sun has ceased to smile. But seed-gathering went very well; in one fold of the Alps a Celestial Poppy gave us pods of so special a size, shape, wartiness and thorniness, containing seed so brown and fat and big and square that I almost dared diagnose a different species . . . The only new treasure of any moment lived in the very highest scree-slopes of all, where every other form of life had faded out in the universal deadness and ruin of Axeinos and Thanatos. This was an Aconite that sat dumpily in voluminous masses of very

handsome foliage scalloped and glossy, close over which emerged
stocky masses of very large blossoms in a close buzzle of helmets, in
texture like Japanese silk-crêpe, and of the oddest, most subtle and
lovely colour, like oyster-coloured chiffon over a blue slip, so that its
effect was that of a delicate French grey or smoke. This strange and
sinister beauty had a tap-root that seemed anxious to get down again
to the bottom of everything; however, with much delving, I worried
up a few of the smaller plants, in a faint hope against hope that I might
get them home alive; but Alas! it was not to be . . .

Round the far bay the circuit of scree grows milder, and so you
come to the end of the Clear Lake, where the real stream brims over
the lip of its moorland barrier in a delicious stretch of clear bubbling
pools and whirls, rippling darkly round among the dark boulders that
lie sunken and becalmed over the face of the fell and its shallow dip.
And the near marge is green alpine lawn again, all along, velvety and
elastic, with the green water lapping lazily against a hem of white silt,
in which springs a new tall Aconite of no particular attraction. Ahead
of you, over the dwarfed chaos of boulders from which the indiscern-
ible minuteness of the camp dominates the flat of the Jardin beneath,
now towers the lumpish massiveness of Axeinos; while, with any luck,
as you serenely wander close along the shore you may get the entire
pyramid of Croda Rossa mirrored in the unflawed emerald of the lake.

But now the harvest was all gathered, and with our burden of
bulging packets it was time to break up the camp and get down again
to Wolvesden to prepare our final departure . . .

And hardly had I started when, in the fine turf that crowned the top
of a sloping boulder, there stared at me a new Gentian, a Gentian that
instantly obliterates all others of its race, and sinks even G. Verna and
G. Gentianella into a common depth of dullness. When the first awe
was over, I gave tongue for Bill, and together, in reverend silence, we
contemplated that marvel of luminous loveliness. Not the faintest
hope possessed me that this glaring miracle could be a new species . . .

The collector's dream is to have some illustrious plant to bear his
name immortal through the gardens of future generations, long after
he himself shall have become dust of their paths. Mere beauty will not

do it; for the plant may fail and fade in cultivation, and his name be no more known, except to the learned, as attached to a dead dry sliver on the sheets of a herbarium. To become vividly immortal in the Valhalla of gardeners, one must own a species as vigorous as it is glorious, a thing capable of becoming, and remaining, a household word among English enthusiasts . . .

Gentiana Farreri . . . bids fair to be as solid a permanency as G. Gentianella itself. It is perfectly hardy, and – what is very remarkable in any Gentian but miraculously so in a Gentian so miraculously beautiful – it is perfectly vigorous and easy to deal with in any reasonable conditions of culture in a cool place not parched or water-logged. Here, indeed, it forms masses many times the size of any clump you will see on its own alps; and already special pilgrimages go to Edinburgh in August and September, to see those jungles of my Gentian, a yard through, with some three hundred gigantic trumpets opening at once. Shall I add that, in addition to growing so freely, and flowering so lavishly in so late and dull a moment of the year, this preposterously good-tempered exception to the rule of its race keeps its glory open, rain or shine, can be struck from cuttings as copiously as a Viola, and layered along its shoots as complacently as any carnation.

And its beauty! Nothing could I foretell of its temper and future history that day, as I stood rapt in contemplation before the actual plant, the last and greatest event of my second season, and well worth the whole two years' expedition anyhow, merely to have seen it. A fine frail tuft like grass radiating some half a dozen fine flapping stems – that is G. Farreri, quite inconspicuous and obscure in all the high lawns of the Da-Tung, even down into the Dene as low as Wolvesden House. Until it flowers; and every day in early September brings a fresh crashing explosion of colour in the fold of the lawns. For each of those weakly stems concludes in one enormous upturned trumpet, more gorgeous than anything attained by G. Gentianella, but in the same general style and form. But the outline is different, with a more subtle swell to the chalice, and that is freaked outside in heavy lines of black-purple that divide long vandykes of dim periwinkle blue with panels of Nankeen buff between; inside the tube and throat are white, but the mouth and the wide bold flanges are of so luminous and intense a light azure that one blossom of it will blaze out at you among the grass on the other side of the valley. In no other plant, except perhaps, Ipomœa Learii, or Nemophila, do I know such a shattering acuteness of colour: it is like a clear sky soon after sunrise, shrill and translucent, as if it had a light inside. It literally burns in the alpine turf like an electric jewel, an incandescent turquoise.

REGINALD FARRER, 1921

When C. M. McKimmon of Hartsville called me to account for saying that camellian is pronounced 'chameleon', I hastened to assure him that it was not pronounced that way by me, and that I spent many years pointing out to others that the 'e' is short.

Those years were spent in vain. I don't think I made a single convert. But he says that ever since Dr Hume told them in a lecture at Coker College that the genus is named for Father Kamel, and pronounced the same way, 'everyone in this section follows his advice'.

I am afraid Dr Hume didn't make such an impression on us, when he talked to the Charlotte Garden Club, for I still hear much of 'cameelias'. When I was young, and a crusader, I set out to reform the garden clubs, but now I am perfectly willing to allow them to call a crinum a lily, a trumpet daffodil a jonquil, a snow flake a snow drop, a camellia a 'cameelia' – or even a japonica, though that does get rather confusing if Japan quinces are called japonicas, too. I even like the flavor of this local usage.

For myself, although I try to pronounce the names of plants correctly, I find it very hard to be consistent. Why make a point of 'kamelia' when you say 'dootsia'? Doesn't Johan van der Deutz deserve as much honor as Father Kamel, and if he does we must say 'doytsia'. And if we love and honor Dr Dahl we must call his name-sake a 'dall'-ya' – but I can't ever remember having heard anyone do it.

Many eyebrows were raised when an airy flower-arranger talked to the Charlotte Garden Club about 'Gerbeerias', but when I got home and looked it up, I found that it was originally spelled and pronounced that way . . .

Once you start trying to do the right thing by plant names, you find that there are a great many that you have mispronounced all your life. Then your lot is not a happy one. For you must choose between continuing in the wrong and feeling very foolish. Having always said 'ox-alice' and 'pitt'-osporum', I cannot now change to 'ox'-alis' and 'pittos'-porum' without a blush or a stammer.

When talking to Lady Whitby I managed to say 'siklamen' twice, but when it came into the conversation for a third time, I broke down completely and went back to 'sighclamen'. I often wonder why you must say 'siklamen' when you say 'sighdonia', but that's the way of it. I do my best, but 'gladeye'-olus' I cannot say. Occasionally there is a choice between the correct and the accepted, but I have never found any justification for Amy Lowell's rhyming clematis with window lattice.

The 'ch' in a plant name is another trap for the unwary. It seems natural to call the wintersweet 'kimonanthus', but it is hard for me to call the turtlehead 'kelone'; and I was undone when I discovered that all these years I should have been saying 'ankusa'. Once I heard a three-year-old correct a visitor who said she admired our lilies: 'They are not lilies,' he said. 'They are "col-ki-cum".' I was about to correct him, but thought better of it, and when I looked it up I found that 'colkicum' is correct, though 'colchicum' is accepted – at least it is accepted in America.

Usage differs when a plant is named for a person. One may say 'Golden'-eye', and another 'Goldeen'-ee'. I say 'Golden'-ee'. Once when I admired a daylily in Miss Nanny Holding's garden in Wake Forest, she said it was *Hemerocallis* 'Golden-eye'. I looked at the wide-eyed golden flower, and thought that a delightful and appropriate name. It never occurred to me to check it, and so *Hemerocallis goldenii* went into *A Southern Garden* as Golden Eye. And as I described it in glowing terms, I expect a lot of gardeners are still searching for it . . .

I used to have a collection of picturesque pronunciations. The only ones I remember now are the 'wiggly rose', which turned out to be *Weigela rosea* (and by the way, if you are going to make a fuss about 'camell'-ia' you should call it vigela); and the 'paleeda delmeetisha', which proved to be *Iris pallida dalmatica*. And of course we commonly meet 'cotton-easter' for 'Coton'-e-aster'. It is all very confusing, and sometimes I feel for the woman who asked if I could tell her the difference between an arbor-vitae and an evergreen.

ELIZABETH LAWRENCE, 1965

I would love to know how old our wisteria is. The trunk certainly looks venerable (I try not to think what the root system may be up to in the foundations . . .). We have a postcard – one of a series of 'Village Views' – which to judge from the children's clothes dates from the teens of this century, and it shows a well-established network of *something*, not in leaf or flower, reaching to the bedroom windows; from the *something's* sinuosity, it could be wisteria. And Albert, who came to the village as a young married man in the twenties, said it looked an old plant then. So it might be a hundred – or more. Albert didn't make a hundred; he was in his eighties when he died, in the spring, one of the last of the true village people. From his parents' garden a couple of miles away he brought clematis cuttings and rose grafts of plants his father had grown from cuttings and grafts. What I saw in his garden were not all the originals, of course, but often 'Son (or "Daughter") of . . .' I remember *Clematis* 'Lasurstern' in mounds and swags of purple velvet, and roses which from their shape and fruity, tarry scent I would guess to have been early Hybrid Teas. He also loved the Scotch thistle, *Onopordon acanthium*, and most years gave me some of the seedlings which came up regularly, though not troublesomely, in his own garden. Oddly, although they throve here and grew into handsome plants, flowered and were left to seed, never did I find a seedling (and no one who knows me will think this was a consequence of over-zealous weeding). But this year I see there is *one*. It has chosen hard and stony ground at the side of a path, which I should not have thought its tap-root would like; I hope it knows what it is doing. If not, there are always the sad purply poppies and giant two-foot nigella in a good blue to remember Albert by.

LIZ ROBINSON, 1991

The birds of this garden are such inveterate berry-thieves, that the amount of pleasure we derive from the consummated efforts of our berry-bearing plants is measured by the extent of avian charity and the share of a crop left for our enjoyment, after our feathered friends have taken all they want. They often strip off the Rowans before they have coloured to their full scarlet, and the fairy raspberries borne so sparingly by *Rubus arcticus* descend into the maw of some Saxifrage-destroying Blackbird, just as though they were the commonest of fruits.

As one cannot preach honesty to the birds, nor would wish to banish their songs and cheerful presence from the garden, it is best to philosophize mildly, and reconcile oneself to the loss of gaily coloured fruits by reflecting that as an actual fact we owe the beauty of colouring of bright berries as much to the birds themselves as to the plants that bear them. The very *raison d'être* of attractive colouring in a berry is to advertise it as a desirable meal for a bird, in order that, in one way or another, the seeds in the fleshy pulp may be carried into suitable places for their germination. But all the same, it requires more philosophy than I possess to prevent my feeling distinctly annoyed when I have been watching some precious crop of fruit from day to day, and suddenly find it has totally disappeared. Perhaps it may be the first time the plant has set seeds, and I want them for sowing; or I may never have seen that fruit before, and wish to learn its ways and beauties, and there is but little satisfaction to be derived from the presumption that a family of thrushes must be suffering from repletion, if not a bad pain inside their speckled waistcoats, after such a heavy meal. *Vitis Coignetiae* fruited very freely here in the hot summer of 1911, and the bunches of green grapes hanging from the cross-pieces of the pergola gave promise of a delightful picture if they finally turned purple. I have never found out what colour they ripen to, however, for one day there remained nothing but bare stalks, and in every season since I have found a few seedlings in the beds close by as proof that at any rate the pips had matured.

E. A. BOWLES, 1915

I t is a familiar experience to find one's greatest aesthetic enjoyment not in a deliberately sought pleasure but in something incidental, the by-product of some other activity. When the conscious mind is occupied with some purely practical task or interest one's eyes may be suddenly opened to the beauty of a familiar scene or to some newly unfolded aspect of things, due perhaps to a trick of light or the changes of the seasons. In many gardens deserving the name none the less for having been planted for a practical utilitarian purpose, such experiences are very precious, and the joy taken in beauty of form and colour may be all the keener for its incidental character. In the orchard or fruit plantation the trees, planted in rows and pruned according to a system, are now being anxiously watched with a view to the maturing of a crop. The grower's aim is practical, his efforts are fully engaged in a battle with the hostile forces that all the time threaten to rob him of the results of his labour, so that his delight in the manifold beauty of fruit blossom at this season [in April] is all the more deeply felt since its enjoyment was never his primary purpose.

The beauty of the English apple orchards in spring is of course proverbial, and there are no varieties that are without a measure of charm at this season. Some are surpassingly lovely however, and anyone planning a mixed plantation would do well to have an eye to this secondary aspect of his undertaking. Perhaps the most beautiful of all is the old 'Brownlee's Russet', which has flowers of a very deep clear pink. The abundant pink and carmine blossom of 'Stirling Castle', 'Lord Derby' and 'James Grieve' are very striking in the mass, while the large and perfectly formed flowers of the 'Orleans Reinette' and 'Warner's King' are especially beautiful when seen at close quarters. Pear-blossom has, as a rule, no very striking charm of colour, but its milk-white clouds are an enchanting spectacle in the sunshine of an early spring morning. The powdered snow of the blackthorn is matched a month later by the blossom of the plum, and here a close examination is as rewarding as in the case of the apple-trees. The perfect five-petalled white flowers with their tuft of golden stamens are crowded in multitudes among the black twigs, which often show little else but the emerald buds of the scarcely opened foliage. It is the blossom of an early flowering plum that figures so often and so charmingly in Chinese art, as in the blue-and-white so-called 'hawthorn jars', which were intended to hold New Year's gifts. The blossom of peach and nectarine has the same individual loveliness, with the added delight of its deep rose colour. Less often noticed are the flowers of quince and medlar. In both cases they are much larger than any of the others I have mentioned, those of the quince being delicately flushed with pink, while the large single roses of the medlar, half-hidden in the young foliage, are a soft milk white.

The æsthetic interest of an orchard is not however confined to the blossom period. Those who take pleasure in the indefinable quality called character will recognize among the varieties of apple many outstanding 'personalities'. The great majority of apple-trees are pruned on the same system for shape and spurs, yet how determinedly do most of them assert their own habit of growth! One could endlessly enumerate one's favourites. Not only such stout upstanders as 'Bramley's', 'Blenheim' and its progeny, and the fabulously vigorous 'Warner's King', but the elegant spreading 'Lane's Prince Albert', the slight and delicate 'Margil' and the almost fastigiate 'Christmas Pearmain' and 'Wagener' all insist on taking their own preferred shapes however they may be pruned or even tied to hoops (for such things are done). This aspect of character is of course as well appreciated in winter, but in summer a particular habit is often matched by a special quality in leaf which may give an equally deep satisfaction. Last in this catalogue of the secondary merits of our fruit trees may be mentioned their beauty of foliage colour in autumn. No quality in trees and shrubs is so uncertain, especially while the subjects are still not fully mature. Soil here provides what I believe is an incalculable factor. Most pear-trees when old colour beautifully in autumn; and the russet and gold of 'Durondeau' and the 'Souvenir du Congrès', which are the best in this respect in my own garden, are doubtless only two among a great number all of equal merit.

WILLIAM BOWYER HONEY, 1939

Did you happen to see that the R.H.S. has recently given an award to Cox's Orange Pippin? I wonder if it feels flattered at receiving, at the age of about a hundred, this belated pat-on-the-back; it's as though the National Book League were suddenly to announce, with a great flourish of trumpets, its discovery that *Vanity Fair* was rather a good novel.

WILFRID BLUNT, 1963

The present way and humour of our gardening in England ... seem to have grown into such a vogue, and to have been so mightily improved in three or four and twenty years of His Majesty's [Charles II's] reign, that perhaps few countries are before us, either in the elegance of our gardens, or in the number of our plants; and I believe none equals us in the variety of fruits, which may be justly called good; and from the earliest cherry and strawberry, to the last apples and pears, may furnish every day of the circling year. For the taste and perfection of what we esteem the best, I may truly say, that the French, who have eaten my peaches and grapes at Shene, in no very ill year, have generally concluded, that the last are as good as any they have eaten in France, on this side Fountainbleau; and the first as good as any they have eat in Gascony; I mean those which come from the stone, and are properly called peaches, not those which are hard, and are termed pavies; for those cannot grow in too warm a climate, nor ever be good in a cold; and are better at Madrid, than in Gascony itself: the Italians have agreed, my white figs to be as good as any of that sort in Italy, which is the earlier kind of white fig there; for in the latter kind, and the blue, we cannot come near the warm climates, no more than in the Frontignac or Muscat grape.

SIR WILLIAM TEMPLE, 1685 (pub. 1692)

These were pleasant feelings, and she walked about and indulged them till it was necessary to do as the others did, and collect round the strawberry beds. The whole party were assembled, excepting Frank Churchill, who was expected every moment from Richmond; and Mrs Elton, in all her apparatus of happiness, her large bonnet and her basket, was very ready to lead the way in gathering, accepting, or talking – strawberries, and only strawberries, could now be thought or spoken of. 'The best fruit in England – every body's favourite – always wholesome. These the finest beds and finest sorts. Delightful to gather for one's self – the only way of really enjoying them. Morning decidedly the best time – never tired – every sort good – hautboy infinitely superior – no comparison – the others hardly eatable – hautboys very scarce – Chili preferred – white wood finest flavour of all – price of strawberries in London – abundance about Bristol – Maple Grove – cultivation – beds when to be renewed – gardeners thinking exactly different – no general rule – gardeners never to be put out of their way – delicious fruit – only too rich to be eaten much of – inferior to cherries – currants more refreshing – only objection to gathering strawberries the stooping – glaring sun – tired to death – could bear it no longer – must go and sit in the shade.'

JANE AUSTEN, 1816

It is surely flying in the face of nature to fill the garden with tropical plants, as we are urged to do by the writers on Landscape Gardening, ignoring the entire difference of climate, and the fact that a colour which may look superb in the midst of other strong colours, will look gaudy and vulgar amongst our sober tints, and that a leaf like that of

the yucca, which may be all very well in its own country, is out of scale
and character amidst the modest foliage of English trees.

SIR REGINALD BLOMFIELD, 1892

I can ... by no means approve of that extravagant Fancy of some,
who tell us that a *Fool* is as fit to be the *Gatherer* of a *Sallet* [salad] as a
Wiser Man. Because, say they, one can hardly choose amiss, provided
the Plants be green, young, and tender, where-ever they meet with
them: But sad experience shews, how many fatal Mistakes have been
committed by those who took the deadly *Cicutæ, Hemlocks, Aconits*, &c.
for Garden *Persley*, and *Parsneps*; the *Myrrhis Sylvestris*, or *Cow-Weed*, for
Chaerophilum (Chervil), Thapsia for *Fennel*; the wild *Chondrilla* for *Succory*,
Dogs-Mercury instead of *Spinach*; *Papaver Corniculatum Luteum*, and
horn'd *Poppy* for *Eringo*; *Oenanthe aquatica* for the *Palustral Apium*, and a
world more, whose dire effects have been many times sudden Death,
and the cause of Mortal Accidents to those who have eaten of them
unwittingly: But supposing some of those wild and unknown Plants
should not prove so *deleterious* and unwholsome; yet may others of
them annoy the *Head, Brain*, and *Genus Nervosum*, weaken the *Eyes*,
offend the *Stomach*, affect the *Liver*, torment the *Bowels*, and discover
their malignity in dangerous and dreadful *Symptoms*.

JOHN EVELYN, 1699

Helmet-floure, or the great Monkes-hood, beareth very faire and goodly blew floures in shape like an Helmet; which are so beautifull, that a man would thinke they were of some excellent vertue, but *non est semper fides habenda fronti* [there is no trusting to appearances]. This plant is universally knowne in our London gardens and elsewhere; but naturally it groweth in the mountaines of Rhetia, and in sundry places of the Alps, where you shall find the grasse that groweth round it eaten up with cattell, but no part of the herbe it selfe touched, except by certaine flies, who in such abundant measure swarme about the same that they cover the whole plant: and (which is very straunge) although these flies do with great delight feed hereupon, yet of them there is confected an Antidot or most availeable medicine against the deadly bite of the spider called *Tarantala*, or any other venomous beast whatsoever; yea, an excellent remedy not only against the Aconites, but all other poisons whatsoever. The medicine of the foresaid flies is thus made: Take of the flies which have fed themselves as is above mentioned, in number twentie, of *Aristolochia rotunda*, and bole Armoniack, of each a dram.

Divers of these Wolfs-banes grow in some gardens.

The force and facultie of Wolfs-bane is deadly to man and all kindes of beasts: the same was tried of late in Antwerpe, and is as yet fresh in memorie, by an evident experiment, but most lamentable; for when the leaves hereof were by certaine ignorant persons served up in sallads, all that did eat thereof were presently taken with most cruell symptomes, and so died.

The symptomes that follow those that doe eat of these deadly Herbs are these; their lipps and tongue swell forthwith, their eyes hang out, their thighes are stiffe, and their wits are taken from them, as *Avicen* writes, *lib.* 4. The force of this poison is such, that if the points of darts or arrowes be touched therewith, it brings deadly hurt to those that are wounded with the same.

Against so deadly a poison *Avicen* reckoneth up certain remedies, which help after the poyson is vomited up: and among these he maketh mention of the Mouse (as the copies every where have it) nourished and fed up with *Napellus*, which is altogether an enemie to the poisonsome nature of it, and delivereth him that hath taken it from all perill and danger.

Antonius Guanerius of Pavia, a famous physition in his age, in his treaty of poisons is of opinion, that it is not a mouse which *Avicen* speaketh of, but a fly: for he telleth of a certaine Philosopher who did very carefully and diligently make search after this mouse, and neither could find at any time any mouse, nor the roots of Wolfs-bane gnawn or bitten, as he had read: but in searching he found many flies feeding on the leaves, which the said Philosopher tooke, and made of them an antidote or counterpoison, which hee found to be good and effectuall against other poisons, but especially against the poison of Wolfs-bane.

The composition consisteth of two ounces of *Terra lemnia*, as many of the berries of the Bay tree, and the like weight of Mithridate, 24 of the flies that have taken their repast upon Wolfes-bane, of hony and oile Olive a sufficient quantitie.

The same opinion that *Guanerius* is of, *Pena* and *Lobel* do also hold; who affirme, that there was never seene at any time any mouse feeding thereon, but that there bee flies which resort unto it by swarmes, and feed not only upon the floures, but on the herb also.

There hath bin little heretofore set down concerning the Vertues of Aconites, but much might be said of the hurts that have come hereby, as the wofull experience of the lamentable example at Antwerp yet fresh in memorie, doth declare, as we have said.

JOHN GERARD, 1597

Don't have ferns in your garden. The hart's tongue in the clefts of the rock, the queer things that grow within reach of the spray of the waterfall, these are right in their places. Still more the brake on the woodside, whether in late autumn, when its withered haulm helps out the well-remembered woodland scent, or in spring, when it is thrusting its volutes through last year's waste. But all this is nothing to a garden, and is not to be got out of it; and if you try it you will take away from it all possible romance, the romance of a garden.

The same thing may be said about many plants which are curiosities only, which Nature meant to be grotesque, not beautiful, and which are generally the growth of hot countries where things sprout over-quick and rank. Take note that the strangest of these come from the jungle and the tropical waste, from places where man is not at home, but is an intruder, an enemy. Go to a botanical garden and look at them, and think of those strange places to your heart's content. But don't set them to starve in your smoke-drenched scrap of ground amongst the bricks, for they will be no ornament to it.

WILLIAM MORRIS, 1882

The Fernery belongs to the truly rustic rather than the rural de-partment of gardening. Though ferns are beautiful anywhere, and may suitably adorn the trim border, and mingle with ornaments of formal design, they are more at home, more befitting among tree-stumps, and in boldly designed rock-work, or water-scenery, where they appear in their proper character of wildness and simplicity . . .

In forming rock-work expressly for ferns, it is best to construct a round or square hillock, with a foundation of lime and brick rubbish, and a surface of about a foot of the proper soil, faced with stones or other material to constitute a rock-work. The south border may be bounded by a high wall or border of shrubs to insure the necessary shade. One side, at least, should never see the sun, one should have it winter and summer, while the other two should but occasionally bask in its rays. The requisites of an open-air Fernery are ample space, variety of sunshine and shadow, plenty of moisture, an atmosphere comparatively pure, alternations of slopes, hollows, and acclivities of surface, and good shelter from high winds and frost.

In town localities, it is difficult to establish ferns in the open air, owing to their delicacy of constitution, and impatience of a dry or smoky air. But in the suburbs of London, any of the ferns that are ordinarily grown in the open air will succeed, as we know by experience, and could name some very flourishing fern gardens at distances varying from two and a half to six miles from St Paul's.

Ferns artificially grown, and tended with proper care and skill, exceed very much the beauty of those grown by nature. True, we cannot grow the scene as well as the fern – we cannot have the dark glen, the dank moss-grown cave, the decayed tree trunk, or the crumbling archway of the waterfall. The scenes amid which ferns grow, the lovely secluded spots which they seek out – shy wood-sprites that they are – are the chief charms of the associations they always suggest to us; for they do haunt the greenest and coolest nooks, the most mossy and ancient banks above water-brooks that trickle from unseen founts, in the deep recesses of wild rocky caverns, and under the branching arms of twisted grey-beard oaks and ancestral beeches – spots only discovered by the explorer of woodbine coverts and deep-hidden shades, where, searching for rare beauty, he finds it far excelling his anticipation, and checking his silent footsteps by sights that hold him breathless with surprise. Yet though we cannot have the mountain dells, and creeping thorns, and purple knolls of wild thyme, we may have the emblems of them in our little mural paradise, we may have the ferns to suggest such things, and to keep alive the remembrance of pleasures and of scenes which keeps a coolness in the brain and a

freshness in the heart – breathings of fragrance from the green world
that sweeten the resting-places in the march of life.

SHIRLEY HIBBERD, 1856

I never before knew the full value of trees. My house is entirely
embosomed in high plain trees, with good grass below, and under
them I breakfast, dine, write, read and receive my company. What
would I not give that the trees planted nearest round the house at
Monticello were full grown.

THOMAS JEFFERSON, 1807 (pub. 1987)

If I look up from my desk I can see the water and the valley beyond
framed by the lower branches of the walnut tree. Its leaves reach
out like seven-fingered hands to within a few feet of the studio
window and its higher branches rustle and sway above the red tiles of
the roof. Thirty feet from the doorway the silver-grey, deeply rutted
trunk, which measures ten feet around its girth, rises only twelve feet
above ground-level before spreading out in a fountain of great limbs.
They writhe and twist sixty feet into the air and spread nearly seventy
feet from side to side. It is magnificent. It makes me eternally grateful
to those of our forebears who planned for future generations rather
than for their own. It is certain that whoever planted the seed and
tended the sapling did not live to see it reach maturity, but I have a

feeling that he knew exactly what he was doing and enjoyed great satisfaction in doing it.

When I look at the tree in the dark days of winter, its huge green-black skeleton silhouetted against the ashen sky, or hear its tracery seething in a westerly gale as I lie snug and warm in bed, I wonder who it was that planted this giant for so many generations to enjoy. And in the balmy days of summer when its leaves are overlaid like the breast feathers of a great bird to form high domes of rounded foliage, I wish I could call back this gentle spirit of the past and say: 'This is your tree. Look at it now, for it is gracious beyond words.'

NORMAN THELWELL, 1978

A garden without trees scarcely deserves to be called a garden, and if I had to build a house for myself, with, of course, a garden attached, I have often thought that it would be very delightful to build the house in an old wood, and gradually form the garden by cutting it out of the wood. Such a thing has often been done. The late Mr Halliwell-Phillips built what he called a homely wooden Bungalow in Hollingbury Copse, near Brighton, and made a garden in the copse; but how far the garden was a success I do not know. In a wood, at Wisley, near Weybridge, Mr Wilson has made a wood-garden, which has almost an European reputation; but it is not a garden, it is still a wood, in which a large number of plants are grown most successfully, and, as it is four miles from the house, it certainly is not a home-garden, and I can only think of a garden as an adjunct to a home. In America also it is not uncommon for rich men to build for themselves a large house in an uncleared forest, and to clear the forest only so far as to make it a park in which the house stands. A small portion round the house would be kept as a lawn, but this does not constitute a garden. Trees by themselves will not make a garden, but a garden lacks more than half its proper beauty if without trees; and so I propose to talk of trees, what are and what are not suitable for gardens.

I should lay it down as a strict rule never to plant English forest trees in a garden. If they are there already, there may be good reasons for keeping some of them, but in a garden of limited extent (and it is of such I am speaking) it seems a waste to plant trees which can be had in perfection in the woods and hedgerows outside of the garden, to the exclusion of the many fine exotic trees which can only be grown in gardens. And even in retaining British forest trees that may be on the ground, I think nothing can be said for them unless they are of some special excellence. There is perhaps no grander deciduous tree than the English oak, when it has past its first manhood and is bordering on old age, and it does seem cruel to cut it down. Yet it takes up too much room when in its full grandeur, especially in certain soils. In some soils it grows to a great height, with a clean straight stem, and not a great spread of branches, as in the grand oaks at Bagot's Park, in Stafford-shire, and if it would always grow like that it might be admitted into any garden, but the general habit is wide-spreading, and I have seen an oak at Southgate, near London, of very moderate height, but the branches cover a circle of which the diameter is over 120 ft. There are very few gardens that could admit such a tree.

On no account should an elm be admitted in or near a small garden, though it is hard to exclude a tree which so rapidly takes a beautiful shape, and which in the late autumn puts on such golden tints. But probably no tree takes possession of a large extent of ground so rapidly as this Italian stranger, for it is not a true native. It sends up suckers at a considerable distance from the parent tree, and these, if let alone, soon become trees. About forty years ago a Scotch forester surveyed a part of South Devon to report on the growth of hedgerow timber, and he reported that in the parish of Clyst St George the hedgerow elms occupied three-quarters of the parish, for though only, or chiefly, in the hedgerows, their roots extended so far into the fields that they met and even overlapped in the middle of the fields, so that only one-quarter of the parish could be considered free from the elm roots.

Beeches must be excluded, for nothing will grow under or near them, and though the copper beech is much admired by many, I could never like it. There are many different shades of copper beeches, of which some may be less ugly than others, and for a few days in the spring the colour of the young leaves is very brilliant, quite equal to the Japanese maples, but a black tree is to my eyes a monstrosity, and sufficiently ugly to justify Wordsworth's complaint, that there were only two blots in his beautiful vale, a copper beech and Miss Martineau. There is, however, one beech, the fern-leaved, which makes a beautiful lawn tree, and has lovely tints both in spring and autumn.

I dismiss at once the horse-chestnut, the sycamore (though, for a few days, when the flowers come out before the leaves, the whole tree is of a rich gold colour), the lime, the plane, and even the Spanish chestnut and the ash, as all better outside the garden than inside; but I would admit one upright poplar (and not more), for its unlikeness to everything around it, and a birch for its exceeding lightness and pretty bark. Of the other native deciduous trees I should be inclined to admit only the hornbeam. Though little grown as an ornamental tree, it makes, when not disturbed or clipped, a very beautiful lawn tree, with a very wide spread of branches which give a pleasant shade, the foliage not being too thick to admit glimpses of sun and light. The best sort is the hop hornbeam, with curious fruit exactly like hops, which are very ornamental, and remain on the tree a long time.

I am not fond of conifers in any part of a small garden; they are for the most part too formal in shape, too thick to give a pleasant shade, and too monotonous in colour. In a park or in large plantations they are very valuable, and in such places I can admire even the *Araucaria excelsa* (the puzzle-monkey), which when seen too near is the most artificial-looking tree in nature, the leaves and branches having almost a cast-iron texture. But there are two evergreen conifers which may be welcomed anywhere. The cedar of Lebanon is one, which (quite apart from its many associations, Biblical and otherwise) is the most delightful tree to grow on a lawn, and if it is in good soil it very soon takes a good shape, so that if I was limited to one tree I should choose a cedar of Lebanon. The deodar of the Himalayas, and the Atlas cedar are probably only geographical varieties, and are fine trees, but not equal to the cedar of Lebanon. The second evergreen conifer which I would not willingly be without is our own British yew-tree. I am not sure that I should plant one, for its growth is so very slow that it will scarcely give *nepotibus umbram*, and until it gets to a good age there is not much beauty in it; but I could scarcely cut down an old one. I am happy in having two old ones on my lawn growing near together, and far beyond the memory of the oldest inhabitant they have carried a swing, and it is pleasant to think to how many generations of the children of the village these yew-trees with their swing have been a never-failing delight. They are represented in an old painting quite two hundred years old. Besides these two evergreen conifers there are two deciduous ones, of which single specimens might claim a place. The larch, brought into England from the European Alps about two hundred and fifty years ago, is very beautiful in spring, and is almost the first of the deciduous trees that comes into leaf, and an old larch covered with lichen is a pretty sight; it is also one of the fastest-growing trees we have, and will grow anywhere. The *Salisburia*, or jingko-tree, from Japan (a near ally of the yew), is another deciduous conifer that should be on every lawn. It is very slow of growth in its earlier years, but its foliage (so like the maidenhair fern) is always pretty and interesting, and in a fine, dry autumn the autumnal tints are magnificent.

I think we are all too shy of planting fruit-trees on our lawns. It is not easy to say why we should not plant apples, which bear (especially the pippins) such lovely flowers in great abundance, and which are equally if not more beautiful when covered with their fruit in autumn; as a summer tree there are certainly many more beautiful. The cherry-tree also is a delightful tree both in flower and fruit, and the autumnal tints are very rich, though not so rich in the cultivated as in the wild cherries, which in some counties, especially in Oxfordshire, are marked features in the woods and on some of the village commons, and in a mild autumn the leaves cling to the trees for a long time. For a short time in the spring the almond has a special beauty, but the flowers are very short-lived, and the tree is more suitable for a shrubbery than for a lawn.

But I must speak more of the many fine exotic flowering trees, which I said should be planted on our lawns in preference to our British forest trees. I suppose there is really no more beautiful flowering tree than the horse-chestnut, but I should not admit it into the garden; for though it has only been introduced about two hundred years, it has become one of our commonest hedgerow trees, and may be admired there; and although the foliage is very grand, the general outline and growth of the tree is too heavy and cumbersome for a lawn. But of all exotic flowering trees there are none to equal the magnolias. We generally grow them as shrubs, or against a wall, but in many places they will grow, and not at all slowly, into fine trees. I remember one magnificent magnolia (I believe *M. grandiflora*) standing alone, and as large as a fine elm, near the old Roman Villa at Brading, in the Isle of Wight. This may be considered a favourable spot, but at Edgbaston Botanic Gardens, near Birmingham, which are in a very exposed and cold situation, there are some grand magnolias which may well take rank as forest trees. They are not the large-flowered species (*M. grandiflora*), but *M. acuminata*, *M. macrophylla*, *M. auriculata*, and *M. purpurea*, and the finest specimen of *M. acuminata* is nearly 50 ft high, and 35 ft through. The tulip-tree (*Liriodendron tulipifera*), from North America, is botanically closely allied to the magnolia, and is a most excellent lawn tree. Its large, quaintly shaped leaves, which in autumn turn to a rich yellow and brown, and its handsome, sweet-scented flowers, make it very attractive; it is also a rapid grower, and there are many trees in England from 100 ft to 140 ft high. The *Catalpa syringæfolia* is another beautiful tree for a lawn, allied to the bignonia, but with trusses of beautiful white and purple flowers, which are unlike any other flower. The drawback to it as a lawn tree is that it is so short a time in leaf. In a backward season the leaves will not appear till the end of June, or even the beginning of July, and will fall with the first frost; but as long as they last they are large and handsome, and the habit of the tree makes it a good tree for shade. The fruit is not often produced in England, but it is very curious, like a long French bean. I had abundance of seed in the Jubilee year, but never before or since; and in the same year the *Kohlreuteria*, from Japan, was also covered with handsome golden fruit. If this tree always produced its fruit I should recommend it, but it will only do so in such exceptional years, and so I do not recommend it for a small collection. Somewhat similar in foliage and even in flower to the catalpa (though not botanically allied) is the *Paulownia imperialis*, from Japan; but not so much to be recommended as the catalpa, because though a magnificent tree with beautiful purple flowers like a foxglove, the flowers come out before the leaves, and are seldom produced at all except in the mildest parts of England. For this reason I grow it as a shrub, or rather as a herbaceous plant, cutting it down to the ground every autumn. Under this treatment very strong shoots are produced in the spring, which carry throughout the summer immense leaves (I have measured them two feet across) of a very delicate colour and texture, so making one of the handsomest foliage shrubs I know.

It would be easy to make the list much larger, but I have in my mind a garden of a small extent, and for such the trees I have named would be almost sufficient. I do not mention such beautiful trees as thorns of many kinds, hollies or service trees, because I should rather rank them as tall shrubs than as trees. But before I close the list I must mention two great favourites, which should be on every lawn, the medlar and the mulberry. I never saw a medlar that was not of a beautiful shape, and it makes a more natural tent or arbour than any other tree; the flower is handsome, and the fruit very acceptable to those who like it. Of the mulberry-tree we cannot say too much. The flowers are inconspicuous, but curious and well worth studying, and the fruit is delicious when we have a hot summer, and the tree is beautiful in shape and colour. An old mulberry-tree is an ornament to a lawn that any owner may be proud of, and it is so easily grown that large limbs cut off and stuck into the ground will grow. Among old gardeners it bore a very high character . . .

Some may think that of the trees I have recommended a few only are suitable for all parts of England, while others are only fit for the warm sheltered situations of the south. But I believe they are all perfectly hardy anywhere if planted in proper situations, and a little attended to when young. I fancy that our forefathers were wiser than we are in their choice of situation for tender trees and shrubs. We are in the habit of putting them in the warmest corners we can find, while they chose the coldest. Leonard Mascall was a very practical gardener, and in 1590 he published *A Booke of the Arte and Manner, How to Plant and Graffe all Sorts of Trees*, etc., and this is his advice: —

Commonly the most part of trees doe love Sunne at Noone, and yet the South Winde (or *vent d'aual*) is very contrary against their nature, and specially the almon tree, the Abricote, the Mulberie, the Figge tree, the Pomegranate tree.

I am sure there is much in this. It is quite certain that all Japanese trees like shade and a north aspect; and the finest and most fruitful old mulberry-tree that I have ever seen is at Rochester, growing in a corner where it looks to the north and east, and is thoroughly protected from the south and west.

CANON HENRY ELLACOMBE, 1895

Men seldom plant trees till they begin to be Wise, that is, till they grow Old and find by Experience the Prudence and Necessity of it. When Ulysses, after a ten years' Absence, was return'd from Troy, and coming home, found his aged Father in the field planting Trees, he asked him, why (being now so far advanc'd in years) he would put himself to the Fatigue and Labour of Planting that which he was never likely to enjoy the Fruits of? The good old man (taking him for a Stranger) gently reply'd: I plant (says he) against my Son Ulysses comes home.

JOHN EVELYN, 1664

The strange and happy coincidence which sent our gardener Mori to us just at this time [1905], with a letter of introduction from Professor Tomari, was of real benefit to the cherries. Mori cleared places here and there among the cedars and made what he called a 'sakura-no' or field of cherries, and later a 'sakura-michi', cherry path. Mori also translated the Japanese names into English; there were, for example, the Tiger's Tail, the Milky Way, and The Royal Carriage Turns Again to Look and See.

Curiously enough, there is an important psychological difficulty which attaches to the name 'cherry' tree in countries like ours [the United States of America] where the cherry is a popular and abundant fruit. This was brought home to me by a question fired at me by Franklin K. Lane, when he was in the Cabinet of President Wilson. He arrived when I was showing our cherry trees to the Japanese Ambassador and his staff, and, with his usual brusque joviality, Mr Lane remarked, 'Cherry trees? Cherry trees? Do they produce good cherries?' In defense of our beloved trees, I retorted, 'Must a rose or dogwood produce a fruit for us to eat?' Marian tactfully suggested that if we called them 'cherry-blossom trees' we would avoid much of this misunderstanding, and I think she is right.

DAVID FAIRCHILD, 1939

The Ilex or Ever-green Oak, is a Tree that deservedly stands in Front of all the Ever-Greens, both for its Beauty and Usefulness. But because there have been already so many Mistakes made, and so many Disappointments undergone with respect to the Kind and Management of this Tree, I have the Pleasure to let the World know from whence those Mistakes came, and how to prevent future Disappointments. The Secret I had from a Friend, who was well informed thereof by a Correspondent in Italy, who assures us, that there are *two Sorts* of Ilex, viz. the Tree, and the Dwarf-Ilex; both of them well known in Italy, by that distinction. And altho' they are exactly alike in Shape and Colour, when they are young, yet it is well known that the Tree-Ilex, will not easily be made a Dwarf, and the Dwarf-Ilex, can never be made a Tree, but is always intended and kept for Espaliers. It is the last of these whose Seed we commonly receive from Italy and Spain; and there have been very few of the former sort ever yet sent over, as imagining that we want Espaliers, and not Timber. Mr Balle in Devonshire, and some few others have been so fortunate as to light upon the large sort, and manage even that according to the Rules of Art, by either letting them stand in the Seed-Plot, without a remove, or if they are remov'd, to do it with the utmost Care and Caution, not hurting or shortning the Tap-root: But as far as I can find, it is the general Complaint, that the Ilex here proves a Dwarf-Tree: And for what reason, let it be judg'd by what hath been said above.

However, after such a Caution as this, it is easie to direct that the Seed be pluck't from the large Trees, and not from Dwarfs, when a demand is made, or to get them from some of those large Trees now in Devonshire, raised from the Acorns, set in well sifted Loam; and not removed if possible, but with all the Root and Earth about them. Thus managed and rightly chosen, it is a Tree of very quick growth, vastly beautiful and very profitable. There is a great deal of the Timber brought over every year into England, the Ship-Carpenters thinking it rather better and tougher than the English Oak. It delights in a deep Soil, rather moist than too dry, and is raised only from Acorns, tho' I am apt to think it may well enough be grafted on a young vigorous English Oak. The Wood of this Tree is sometimes finely Chamletted, as if it were painted. It is useful for Chairs, Axle-Trees, &c. being very hard and durable. And the Acorns are thought a Food for Hogs little inferior to Chesnuts.

JOHN LAURENCE, 1726

I have seene at Cobham in Kent, a tall or great bodied Lime tree, bare without boughes for eight foote high, and then the branches were spread round about so orderly, as if it were done by art, and brought to compasse that middle Arbour: And from those boughes the body was bare againe for eight or nine foote (wherein might be placed halfe an hundred men at the least, as there might be likewise in that underneath this) and then another rowe of branches to encompasse a third Arbour with stayres made for the purpose to this and that underneath it: upon the boughes were laid boards to tread upon, which was the goodliest spectacle mine eyes ever beheld for one tree to carry.

JOHN PARKINSON, 1629

From time to time a plant is adopted by a cult or religion either as a symbol or as an object of veneration in itself and, as such plants must have some peculiar or outstanding quality to be thus chosen, they are worth regarding and perhaps cultivating, even apart from the faint halo of romance with which they have been endowed by their associations.

The Mistletoe is the only obvious cult plant which is native to this country and it is not surprising that it should have been chosen or that it should have retained its distinction, for it looks more like some olive-coloured seaweed set with pearls than a land plant, and its peculiar habitat, midway between earth and sky, easily suggested that its origin was supernatural. Its association with the Oak is puzzling, for you will not find more than two or three oak trees in the country today with Mistletoe growing on them; yet evidence for the association does not depend entirely on Pliny's rather supercilious account of the Druids, for in Anglo-Saxon our Mistletoe was known as 'oak-mistel' to distinguish it from Basil, which was known as 'earth-mistel' – it is impossible even to guess at the train of thought by which two such dissimilar plants were linked together. The cult of the Mistletoe has survived the old religion and it is still regarded with slight disfavour by the new, for it is, I believe, the only plant which is banned from church decorations; but it has come down in the world and its strangeness is somewhat dimmed for us by its part in the half-hearted saturnalia of Christmas.

JASON HILL, 1940

We had begun to focus most of our energies on shrubs long before the war came to step up that need for economy, which falling dividends, rising taxation and increasing labour difficulties, had made imperative. But the change came gradually, if, indeed, change it was, for shrubs had ever been our main issue. It would be nearer the truth to say that our never very exalted efforts in the way of herbaceous borders, our attempts to grow alpines in a humid tree-belted glen, came to an end in a piecemeal fashion. The more troublesome herbaceous things made way for shrubs, as room for the latter became needful, and, as for the rock-plants, these, subjected to that attention which speedily follows absence in replacement, soon became submerged . . .

To all but single-handed gardeners, like ourselves, our early concentration upon shrubs was largely the offspring of circumstances other than those mentioned. True, like not a few, we had felt the pinch of economic conditions, but it was the problem as to how to maintain the garden by our own labours that was the snag to be overcome. We might, of course, have increased our help from a single part-time man, and, as a matter of fact, did do so for considerable periods, especially during the early days of preparation of this or that newly acquired addition to the premises. But it was never our ambition to have our work done by others. We enjoyed doing it with our own hands, putting our own selves into the job and, for better or worse, developing the garden after our own desires, often to reap abiding satisfaction, sometimes to endure punishment condign.

That the growing of shrubs absolves one from all toil and trouble I am not going to pretend. But that it does substantially cut-down expenditure in time and money cannot be controverted. Indeed, one might go so far as to say that an established shrub border, in demanding practically nothing in the way of upkeep, is perhaps the most economic of all phases of gardening, that is if you are not too fastidious. Thus, some of our older shrub plantations, which now rarely get more than a few hours' attention in the course of the year, might shock the unfortunate possessor of a tidy mind. For these shrubs, once interplanted with innocuous ground covering plants, bulbs and other plants, now have an undergrowth that may be anything but disciplined, but which is at all seasons pleasing to look upon. It is more or less wild and if it happily renders hoeing impossible it smothers most weeds, only occasionally appealing to us to remove a stray bramble or colony of creeping buttercup. This aspect of our proceedings holds so important a place in our garden maintenance I shall return to it as occasion offers.

But, comforting as such borders and beds are in their independence, we were, with that extension of the garden proceeding every few years, ever confronted by newly planted areas which necessarily demand care. While the garden, with the ever-ready help of nature, does its best to cover our progress and rub in the lesson that maturity is our most desired goal and only salvation, ambition ever urges us to fresh ventures. And this when we had already as much and often more than we could accomplish . . .

To dwell at this date in a general way upon the charms and attributes of shrubs, including the smaller trees, would surely be superfluous, for while the economic advantages of shrubs compared with herbaceous plants have been amply proved, their garden value from an entirely aesthetic point of view is manifest, not least being that year-round interest they provide. In no other department of gardening is that all-season provision of colour and form so pronounced as that in which shrubs and trees prevail. None other does so much in so brief a period to endow the garden with that grace of ripeness which is, or should be, the primary objective of each one of us.

Even in early February, often the deadest month of the year, the shrub's part in the garden furnishing (if one must use the horrid term) is manifested in no uncertain way. Winter cherry and witch-hazel, viburnums and bush honeysuckles, jasmine and *Erica carnea*, mahonias and mezereum, garrya and forsythia and rhododendrons, these come through the acid test of mid-winter with a triumph the herbaceous border can never claim, a triumph, too, of horticultural progress, recalling the time when, if the winter garden was recognized at all as a desirable possibility (though Bacon urged its claims), its dominant feature consisted of little more than evergreens of which laurels were all-pervading. Not that we overlook the contribution made by evergreens *per se* at such a season or any other. Indeed we should be disloyal to many of our best friends did we overlook the manifold charm of rhododendron leaves, large and small, the infinite diversity in their forms, their extensive range of colours and individual poise.

The conifers are no less appealing, while the glancing lights twinkling from the little polished leaves of *Azara microphylla* at snowdrop-time, when from those elegantly feathered boughs there comes the sweet vanilla scent cast forth by swarms of secretly concealed blossoms, would, alone, be evidence enough of shrub value estimated at the season of life's lowest ebb.

ARTHUR TYSILIO JOHNSON, 1950

It is the fashion, especially among botanists, to despise variegated plants; they are said to be diseased, and to show their disease by their sickly appearance. This may be true, if the partial absence of chlorophyll is to be counted a disease, but it does not prevent their being very beautiful and very useful, and if vigour in growth is a sign of health, there are many instances in which the variegated specimens are hardier and more vigorous than the typical green forms. But I here claim for them a special value in the winter decoration of our gardens, and will name a few which I find most useful for this purpose. I have seen some of the variegated ivies covering a large extent of wall, and at a very small distance the wall seemed to be clothed with a rich creeper bearing an abundance of yellow or white flowers. All the variegated hollies are as useful; they brighten up a lawn in a wonderful way, and there are many to choose from, but I find that the variety with the large white blotches on the leaves produces the best effect. It is called, I believe, the milkmaid holly, but it has other names; and one peculiarity of it is its habit of bearing branches in which every leaf is of a pure ivory white; this, too, at a little distance looks like a bunch of white flowers, and if cut will keep its beauty in water all through the winter. I have also on a rock-work a large mass of the dwarf Japanese euonymus (*E. radicans variegata*), which I much prize. Not only does it lighten up the rock-work at all times, and especially in the winter, but it gives abundance of pretty sprays which are most useful both in the house and the church.

CANON HENRY ELLACOMBE, 1895

I have discovered rhododendrons, and my gardening friends don't know what to make of it. Rather like an interest in horse racing or having a son called Harry, this is traditionally a province of either the aristocracy or the vulgar classes, rarely of the tasteful middle ranks. We are interested in shrub roses and herbs, in muted colours and 'architectural' things; not flowers the size of footballs in shades of Barbara Cartland pink.

Actually it is not so much the footballs that have fired my imagination (though it is probably only a matter of time before I embrace them too); it is more the open rosettes of funnels displayed by the azaleas and the smaller species and hybrid rhododendrons. These not only provide a range of exciting colours for the spring garden but present themselves with style and elegance. Some also possess a heart-stoppingly sweet scent and leaves that catch fire in the autumn.

Why has it taken me so long to appreciate them? Part of the reason must be that I had never toured the great woodland gardens of Devon and Cornwall at peak flowering time. But last May I did, and my jaw never left the ground. Orange, red and yellow Ghent and Knap Hill azaleas in pools of bluebells; clear pink *Rhododendron vaseyi* – what a charmer! – tucked under the skirts of a vast, fragrant, blush white *R. loderi* 'King George'; *R. falconeri*, with rust-coloured undersides to its leaves, in association with a tan-trunked birch, *Betula ermanii subcordata*, and a tan-flushed fern, *Blechnum penna-marina*; pale pink *R. trichostomum*, so dainty it could be mistaken for a daphne, all alone in a woodland clearing; and white 'Lady Alice Fitzwilliam' in front of evergreens at the corner of a path, bowling over passers-by with its fragrance: these were some of the encounters that made me wonder whether I had been gardening on another planet up until now.

So there is frantic activity in my Welsh garden this season. Room has to be found for the shipments of plants speeding their way to me from nurseries all over rhodoland. Shrub roses, you'd better look out!

STEPHEN LACEY, 1989

The pruning and trimming of all the Ivy walls and festoons has been done. The result for the time is as ugly as it is desirable. Ivy grows so lavishly here that it has to be kept well in hand, and many whom it favours less, have said they envied us our Ivy. More than once we have had to choose between some Tree, or a canopy of Ivy. It is like a beautiful carpet underneath a long row of Elms, where nothing else would grow; indeed, wherever there happen to be bits too over-shadowed for grass or otherwise unsatisfactory, we put in Ivy; it is sure to understand, and to do what is required. My favourite sort is the wild English Ivy, and no other has a right to grow on the House. Its growth is slow, and sure; it always grows to beauty, and never to over-richness. The loveliness of its younger shoots and of the deeply cut leaves might inspire either poet or painter! To either I would say, wherever on your tree, or fence, or house-wall, you find it beginning to spring, cherish it; for years it will do no harm, and if you are true to your art, and therefore know that small things are not too small for you, it will repay your love a hundredfold. Wild Ivy is best where it comes up of itself, it clings then so close and flat.

MRS E. V. BOYLE, 1884

My dear GrandPapa

The tuberoses and Amaryllis are taken up. We shall have a plenty of them for the next year. The tulips and Hyacinths I had planted before I left Monticello. They had increased so much as to fill the beds quite full. The Anemonies and Ranunculus are also doing well. Fourteen of Governor Lewis's Pea ripened which I have saved. The Pinks Carnations Sweet Williams Yellow horned Poppy Ixia Jeffersonia everlasting

Pea Lavetera Columbian Lilly Lobelia Lychnis double blossomed Poppy and Physalis failed, indeed none of the seeds which you got from Mr McMahon came up. Ellen and myself have a fine parcel of little Orange trees for the green house against your return. Mrs Lewis has promised me some seed of the Cypress vine. Mama Aunt Virginia and all the children are well and send their love to you. Good night my dear Grand Papa. Believe me to be your sincerely affectionate Grand daughter.

ANNE CARY RANDOLPH to her grandfather,
Thomas Jefferson, 1807 (pub. 1987)

I scarcely dare trust myself to speak of the weeds. They grow as if the devil was in them. I know a lady, a member of the church, and a very good sort of woman, considering the subject condition of that class, who says that the weeds work on her to that extent, that, in going through her garden, she has the greatest difficulty in keeping the ten commandments in anything like an unfractured condition. I asked her which one? but she said, all of them: one felt like breaking the whole lot. The sort of weed which I most hate (if I can be said to hate anything which grows in my own garden) is the 'pusley', a fat, ground-clinging, spreading, greasy thing, and the most propagatious (it is not my fault if the word is not in the dictionary) plant I know. I saw a Chinaman, who came over with a returned missionary, and pretended to be converted, boil a lot of it in a pot, stir in eggs, and mix and eat it with relish, 'Me likee he.' It will be a good thing to keep the China-men on when they come to do our gardening. I only fear they will cultivate it at the expense of the strawberries and melons. Who can say that other weeds which we despise, may not be the favourite food of some remote people or tribe. We ought to abate our conceit. It is possible that we destroy in our gardens that which is really of most value in some other place. Perhaps, in like manner, our faults and vices are virtues in some remote planet. I cannot see, however, that this thought is of the slightest value to us here, any more than weeds are.

CHARLES DUDLEY WARNER, 1876

Lady Cantal says that flowers can feel pain. I asked her if hers did, and she said yes, so I pointed to a hideous stone gnome overlooking a bed of very nice Stocks and told her the reason.

Luckily Lady Cantal does not take offence easily, and she went on explaining her theory. I did not hear very much of what she said, my attention being riveted on a leaden frog that gaped uncomfortably and looked as if it might be sick at any moment. I had a feeling that if I watched it long enough it would be, and I did not want to miss the performance.

But it only went on gaping.

Still, there is no reason why Lady Cantal should not keep a menagerie of stone and lead in her garden if she wants to. It is her garden. I wish she did not have pet names for each of her horrors, though.

She is really quite a nice woman. When I first knew her, I thought I disliked her very much; but I found, after a time, that the feeling was only annoyance at her patronage. Like a lot of other people, I did not like to think I was not appreciated at my true worth. Then I reflected that perhaps my true worth was the figure at which she had assessed me, and I decided to leave it at that.

Lady Cantal will live to be a centenarian, and on her hundredth birthday all the people she has bossed and interfered with and bullied will come and tell her what a wonderful person she is and how they could never have enjoyed their lives as they have done without her help.

'Interfering old faggot!' says Micah.

She was showing me all round her rock garden one day when a splash of deep rich yellow caught her eye. We went to have a look at it.

She stopped to examine the tiny, pea-shaped blossoms and turned to me with a puzzled frown.

'That's funny,' she said. 'It should be a pink Lewisia here.'

'That certainly isn't a Lewisia,' I pointed out.

'No, I know. Do *you* know what it is?'

I toyed with the idea of fooling her with a Latin name, but remembered that Barleycorn, the gardener, would know for certain. And then she would think I did not.

'It's Bird's-foot Trefoil.'

'Good gracious!' she exclaimed in horror. 'That's a weed, isn't it? I must pull it out at once.'

She was about to do so.

'Why not leave it?' I suggested. 'It looks rather nice.'

It did look nice, too. It was a neat, tight little clump, and was covered with flowers.

She looked at me suspiciously.

'Leave it? A weed?'

I waved an arm vaguely. 'Well, they're all weeds, aren't they, more or less. You grow foreign weeds. Why not give an English one a chance?' I pointed out again. 'It looks very nice.'

'Ye-es.' She sounded doubtful. 'Would *you* have it in your little rock garden?'

Other people's possessions are always 'little'. 'Your little garden', 'Your little cottage', 'Your little stories'.

'I would,' I said. 'Definitely.'

She was still not sure. 'But you grow such *funny* things sometimes.'

'It would be fun growing that. Half the people who saw it wouldn't know what it was and would think it was some rare plant.' That moved her a little. 'It could easily be pulled out if it spread and became a nuisance.'

'I think I will leave it,' she conceded. 'It looks so pretty – and the Lewisia is obviously dead.'

So she left it. And the funny thing is that it has lived on in that pocket — not spreading rapidly, but growing a little year by year — and it flowers marvellously.

Every time I see it I give it a wink and it winks back at me.

No, it does not wink at all, but after being with Lady Cantal an hour one is apt to say things like that.

H. L. V. FLETCHER, 1948

THE FLOWER

Once in a golden hour
  I cast to earth a seed.
Up there came a flower,
  The people said, a weed.

To and fro they went
  Thro' my garden-bower,
And muttering discontent
  Cursed me and my flower.

Then it grew so tall
  It wore a crown of light,
But thieves from o'er the wall
  Stole the seed by night.

Sow'd it far and wide
  By every town and tower,
Till all the people cried,
  'Splendid is the flower.'

Read my little fable:
   He that runs may read.
Most can raise the flowers now,
   For all have got the seed.

And some are pretty enough,
   And some are poor indeed;
And now again the people
   Call it but a weed.

ALFRED, LORD TENNYSON, 1847

It is fashionable nowadays to affect a horror of bedding plants. People say they must allow a few to please the gardener, just as they say they eat entrées and savouries to please the cook. The simple life is becoming an affectation in dinners and gardens; the table-cloth goes, but hard labour in polishing tables falls on the footmen, and even if you dine on a Tudor oak table you have a lace mat, and under that another fandangle, some bad conductor of heat for your plate to sit on. Simple? It reminds me of a silly old song that lilted of someone being as 'simple as Dahlias on Paddington Green'.

So we despise Scarlet Geraniums as we miscall our Zonal Pelargoniums, make a face at Calceolarias, and shudder at the mention of Blue Lobelia. Well, they have been sadly misused, I know, but they are fine plants, for all that, when in their right places. I remember a garden of twenty years ago that was the most bedded out I ever saw. Thousands of bedding plants were prepared for planting out in Summer, but always in straight lines in long, straight borders. It all began at the stable gates, and ran round three sides of the house, and continued in unbroken sequence, like Macbeth's vision of kings, for two sides of a croquet lawn, and then rushed up one side and down the other of a long path starting at right angles from the middle of the lawn, and if you began at the gates with Blue Lobelia, Mrs Pollock Pelargonium, Perilla, Yellow Calceolaria, and some Scarlet Pelargonium in ranks according to their relative stature, so you continued for yards, poles, perches, furlongs, or whatever it was – I hate measures, and purposely forget them – and so you ended up when the border brought you back again to the lawn. I once suggested, Why not paint the ground in stripes, and have the effect all the year round, even if snow had to be swept off sometimes?

E. A. BOWLES, 1914

With flowers, as with all other departments of the garden, you first decide what kind you *want* to grow and then whittle it down to what kind you *can* grow in the space. When you have further gone through the list and reduced it to what you can *afford* to grow and then eliminated the ones which you know from bitter experience will *refuse* to grow you have saved yourself a very great deal of labour indeed.

ETHELIND FEARON, 1952

The only garden flowers I care for are the quite old-fashioned roses, sunflowers, hollyhocks, lilies and so on, and these I like to see growing as much as possible as if they were wild. Trim and symmetrical beds are my abhorrence, and most of the flowers which are put into them – hybrids with some grotesque name – Jonesia, Snooksia, – hurt my eyes. On the other hand, a garden is a garden, and I would not try to introduce into it the flowers which are my solace in lanes and fields. Foxgloves, for instance – it would pain me to see them thus transplanted.

GEORGE GISSING, 1903

In America 'ghost gardens' are rather the fashion. Here everything is dim and subdued. Under cool, creeper-clad pergolas or shady trees are arranged such misty effects as can be got by the use of large masses of *Gypsophila paniculata*, the great silver thistle, tall grey mulleins, and of the broad-growing silver salvia or sage as a flat carpet from which rise fragile white campanulas, white moon-daisies, or white lilies of Bermuda. Near a grey stone seat may be found a large clump of white datura, with heavily scented trumpet flowers; or, if no frost-proof shed can be found to store the datura, then a group of white *Nicotiana sylvestris* or *Nicotiana affinis*, which comes up year after year. Then there are white foxgloves, with white jasmine as a background, on some old wall or pillar; and *Clematis flammula* (or, earlier in the season, *montana*) wreaths itself in and out of bay or box, and even tall junipers standing like sentinels do not escape, but are caught in its embrace. A small marble basin with water-lilies growing in it is sunk in the cool green turf, the water reflecting the early moonbeams; for no one walks in the ghost garden except at evening.

ALICE MARTINEAU, 1923

The subject of Auriculas is one of such fascination that it is nearly impossible to know how to begin. Perhaps the only sensible method is to describe a personal experience of these flowers. This will betray my ignorance, only a few years ago. But that does not matter. Ignorance is nothing, where there is the will to learn. Auriculas, or Bears' Ears, were plants I had seen growing in the corner of a neglected, weed-grown bed, in one of the old gardens at my home in Derbyshire. That particular garden had been laid out, as I knew, by my great-grandfather soon after his marriage, early, in fact, in the eighteen twenties. And I knew that little, or nothing, had been done to it since 1846, when he went to live abroad. The plants were, then, some seventy or eighty years old when I first remembered them. As I see them, now, they were plain sterling Auriculas of the Alpine type, such as would be seen growing in any garden of that day. They carried with them, as was natural, some hint of the pre-Victorian era. Their colours, as I remember them, were a yellow, a maroon, and a blue of the type of 'Old Irish Blue'. . .

Only some four or five years ago one of the spring shows at the Royal Horticultural Society's Hall – this was towards the end of April – had a display, at one end, of two stands from the firms of James Douglas and of G.H. Dalrymple. Auriculas, in fact, from the only comprehensive growers of these flowers. It was for the first time that I beheld a Stage Auricula. I may, therefore, be able to describe the impression of this from a point of view that has not become dulled by long familiarity. In the first place, it is hardly to be conceived possible that such a plant should exist at all. There is something unreal and improbable in its edging and mealing. This latter has been stippled or dappled on to it. The white, mealy eye of the flower is a glorious and wonderful thing; but the slight and miraculous powdering upon the back of the flower is even more striking. Then the meal upon the leaves of the plant is an æsthetic or even sensual pleasure of the highest order. And the perfect corona of so many heads in flower, the cluster, or truss, has the plenitude, the richness, of a Bacchic bunch of grapes. But we must revert again to the mealing or powdering upon the flower. A simile for this can only be found in the language of another epoch. We find it in mediæval terms of description, where it is to be read, for instance, in the accounts for the great wardrobe of Edward III, that he possessed a bed of red worsted powdered with silver bottles having tawny bands and curtains of sindon beaten with white bottles; or, in another inventory, of three curtains of tartyrn beaten with garters to match, and a gown of blue cloth of gold of Cyprus powdered with golden stags. It is those technical terms, 'powdered' and 'beaten', that are of application to our present purpose. No other phrases can be found to indicate this effect of stippling, which is laid upon the leaves and petals as though in individual touches from the point of the finest camel's hair brush, or to describe the massed and drilled farina of the plant's white eyes. But it would be a mistake to deduce from this that there is anything in the least mediæval, or of the Gothic age, where the Auricula is concerned. On the contrary, there is no other florist's plant, not even the Tulip, which is so instinct of the years in which it thrived.

This first moment of seeing a Stage Auricula is an experience never to be forgotten. It would seem incredible that a flower, through human skill, should attain to this degree of natural or trained artificiality. For the perfection of a Stage Auricula is that of the most exquisite Meissen porcelain, of the most lovely silk stuffs of Isfahan, which is to say that it attains to the highest technical standards of human craftsmanship. And yet, it is a living and growing thing, with the gift of procreation. It is a matter of taste; but it might be argued that a fine Stage Auricula is among the most beautiful things in nature, or in the

world of human beings. If, that is to say, nature does not necessarily
and always mean a wild and uncontrolled growth, a picturesque confusion. For the Auricula epitomizes nature controlled and in the service
of man. It belongs to epochs of discipline and order, to the day of a
living architecture and a law of life. The effect of this flower is to be
compared to that slight disintegration of the senses when features or
limbs of a ravishing loveliness are seen. A perfection of physical
beauty produces this bewilderment and wonder. It is incredible and
cannot be believed. It must be looked at in silence, for fear that it
should vanish. And then the humanity of this flower dawns in comfort
upon the mind; and this is to be compared to the voice, or to the
expression in the eyes, of that human vision, which now reveals itself
to be growing from the soil. Such, then, akin to the sensation of falling
in love, must be our introduction to the Stage Auricula.

SACHEVERELL SITWELL, 1948

The Giant Lily ... has matched her stature against the great fis-
sures and precipices and nameless peaks. In an English garden
she looks startling indeed, but out there a peculiar fitness must attend
her, making of her the worthy and proportionate ornament, sculptural
as she is with her long, quiet trumpets and dark, quiet leaves. I do not
know to what height she will grow in her native home, but in England
she will reach twelve feet without much trouble, and I have heard it
said that in Scotland she will reach eighteen.

A group of these lilies, seen by twilight or moonlight gleaming
under the shadow of a thin wood, is a truly imposing sight. The scent
is overpowering, and seems to be the only expression of life vouch-
safed by these sentinels which have so strange a quality of stillness. I
should like to see them growing among silver birches, whose pale

trunks would accord with the curious greenish-white trumpets of the flower-spike. Unluckily, we have not all got a birch-wood exactly where we want it; and even though we were willing to make a plantation, the stem of the young birch lacks the quality of the old. Alders would do well; they have the requisite pallor and ghostliness.

But failing either of these, any coppice say of hazel or chestnut will serve the purpose, which is to provide shade and coolness, for the Giant Lily will stand a good deal of both. Then you must dig out a hole two to three feet deep, and fill it with the richest material you can provide in the form of leaf-mould, peat, and rotted manure. This simple recommendation reminds me of the exclamation of a friend: 'It seems to me,' she said, 'that this lily of yours has all the virtues and only four disadvantages: it is very expensive to buy; the bulb takes three years before it flowers; after flowering once it dies; and you have to bury a dead horse at the bottom of a pit before it will flower at all.'

VITA SACKVILLE-WEST, 1937

The Madonna Lilies, or rather some brown-looking stalks and a few faded leaves, are over. What is the mystery about this lily? As usual, the stalks looked robust, and healthy enough, and bore thick clusters of pale green buds, but before these buds were able to unfold and reveal themselves to us in their full glory of cream and gold, some deadly blight descended on them and they shrivelled up and withered away.

The bulbs, which have been in the garden for many years now, do very occasionally give us a glimpse of how beautiful a thing they really can be when in full flower; but this they do very seldom. As a rule, just before Christmas, strong green leaves, which bronze slightly as the winter advances, push up through the earth. But directly summer comes, there comes with it this dreaded scourge, and the tender buds, so full of lovely promise, are stricken and laid low.

Every recommendation, from leaving them strictly alone, to moving them continually – besides quantities of Bordeaux mixture, and flowers of sulphur – has been tried, but to no avail.

To the cottage dweller they present no difficulties whatsoever; for in most cottage gardens they flourish and increase, adding several cubits to their stature every year. Large bunches of them can be seen being taken by 'bus or lightly tied to the handles of a bicycle, to be given to distant friends. The bulbs are torn from the ground at all seasons of the year, to be passed on to other cottage gardens, where they quickly settle down and become more beautiful than the polished corners of the Temple. And, as far as can be ascertained, no attention beyond the washing-up water, when it is quite finished with, is given them. And on this mixture of grease, strong soda and soap they are able to surpass Solomon in all his glory.

ETHEL ARMITAGE, 1936

If I had plenty of suitable spaces and could spend more on my garden I would have special regions for many a good plant. As it is, I have to content myself with special gardens for Primroses and for Pæonies and for Michaelmas Daisies. And indeed I am truly thankful to be able to have these; but we garden-lovers are greedy folk, and always want to have more and more and more! I want to have a Rose-garden, and a Tulip-garden, and a Carnation-garden, and a Columbine-garden, and a Fern-garden, and several other kinds of special garden, but if I were able, the first I should make would be a Wall-flower garden.

It should be contrived either in connection with some old walls or, failing these, with some walls or wall-like structures built on purpose. These walls would shock a builder, but would delight a good gardener, for they would present just those conditions most esteemed by wall-loving plants, of crumbling masonry built of half-formed or half-rotting stone, and of loose joints made to receive rather than to repel every drop of welcome rain. Wall-flowers are lime-loving plants, so the stones would be set in a loose bed of pounded mortar-rubbish, and there would be sloping banks, half wall half bank. I should, of course, take care that the lines of the garden should be in suitable relation to other near portions, a matter that could only be determined on the precise spot that might be available.

But for the planting, or rather the sowing of the main spaces, there would be little difficulty. I should first sow a packet of a good strain of blood-red single Wall-flower, spreading it over a large stretch of the space. Then a packet of a good yellow, either the Belvoir or the Bedfont, then the purple, and then one of the newer pale ones that have flowers of a colour between ivory-white and pale buff-yellow. I would keep the sowings in separate but informal drifts, each kind having its share, though not an equal share, of wall and bank and level. Some spaces nearest the eye should be filled with the small spreading Alpine Wall-flowers and their hybrids, but these are best secured from cuttings ... A few other plants would be admitted to the Wall-flower garden, such as yellow Alyssum on sunny banks and Tiarella in cool or half-shady places, and in the wall-joints I would have in fair quantity the beautiful *Corydalis capnoides*, most delicate and lovely of the Fumitories. Leading to the Wall-flower garden I should like to have a way between narrow rock borders or dry walls. These should be planted with Aubrietias, varieties of *A. græca*, of full and light purple colour, double Cuckoo-flower in the two shades of colour, and a good quantity of the grey foliage and tender white bloom of *Cerastium tomentosum*, so common in gardens and yet so seldom well used; I would also have, but more sparingly, the all-pervading *Arabis albida*.

These plants, with the exception of the Cuckoo-flower, are among those most often found in gardens, but it is very rarely that they are used thoughtfully or intelligently, or in such a way as to produce the simple pictorial effect to which they so readily lend themselves. This planting of white and purple colouring I would back with plants or shrubs of dark foliage, and the path should be so directed into the Wall-flower garden, by passing through a turn or a tunnelled arch of Yew or some other dusky growth, that the one is not seen from the other; but so that the eye, attuned to the cold, fresh colouring of the white and purple, should be in the very best state to receive and enjoy the sumptuous splendour of the region beyond. I am not sure that the return journey would not present the more brilliant picture of the two, for I have often observed in passing from warm colouring to cold, that the eye receives a kind of delightful shock of surprise that colour can be so strong and so pure and so altogether satisfying. And in these ways one gets to know how to use colour to the best garden effects. It is a kind of optical gastronomy; this preparation and presentation of food for the eye in arrangements that are both wholesome and agreeable, and in which each course is so designed that it is the best possible preparation for the one next to come.

I think I would also allow some bold patches of tall Tulips in the Wall-flower garden; orange and yellow and brown and purple, for one distinct departure from the form and habit of the main occupants of the garden would give value to both.

GERTRUDE JEKYLL, 1900

One of the first flowers to greet the spring is *Pulmonaria officinalis*, a humble member of the borage family. And how welcome are those jaunty sprays of innocent flowers in blue and pink, often rising above the snow, and fluttering unconcernedly in the bitter winds of March. The common name is Lungwort, from *pulmo*, a lung, as the plant was used in the old days to cure lung disease. But, like other popular plants, that is only one of many names. The little pink and blue flowers growing together reminded someone of the old saying 'pink for a girl and blue for a boy' so we have Boys and Girls as one name, and Soldiers and Sailors for another. In Somerset they like Bloody Butcher, going back to the days when every self-respecting butcher wore a striped blue apron round his portly person. More romantic is Hundreds and Thousands, for a big clump of pulmonaria is a delightful medley of pastel tints. Another old name is Joseph and Mary, why I do not know.

Pulmonaria officinalis was the one grown in most cottage gardens, with heart-shaped, spotted leaves, as rough as a calf's tongue, with blue and pink flowers growing together. The white flowered form is icy and aloof and may be too pure for those who like richness of colour in their flowers.

Pulmonarias are very popular today with our liking for handsome foliage and good ground cover, but connoisseurs are choosy about them. They want the one known as *P. saccharata* because its leaves are longer and more heavily spotted. Collecting good forms is a regular gardening sport, and some of those found are so heavily covered with spots that they are practically silver. There is one good pulmonaria in this family called *P. saccharata* Mrs Moon, with very good foliage and flowers that open pink then turn to blue, but I have never found out who Mrs Moon was or where she lived.

The red Pulmonaria *P. rubra* comes out very early, often before the blue and pink. It is a fine reward for facing the draughty air of a February day to come face to face with little clusters of tight coral flowers tucked into cups of soft green foliage. *Pulmonaria rubra* has no markings on its leaves and the flowers are the brightest shade of pure coral. Sometimes this plant is called Bethlehem Sage, and I have heard country folk refer to it as the Christmas Cowslip. There is a particularly good form known as Mr Bowles' Red. It does not appear to be very different from the ordinary type except that the habit is a little more upright and the flowers a little brighter and bigger, and is just another example of the late Mr Bowles' unerring eye for a good plant.

The blue Pulmonarias are a little higher in the social scale and are sometimes considered worthy of a place in a rock garden. There is a little confusion about the names. *P. angustifolia* appears to be the same as *P. azurea* with *P. angustifolia azurea* as the name for one with light-blue flowers. They divide again into Mawson's Blue and Munstead Blue, but I have never found anyone yet who can find much difference between them. I prefer Munstead Blue, which must have come from Miss Gertrude Jekyll's garden, and is, I think, a more refined plant. It is deciduous and slow-growing. The leaves appear before the flowers as little folded pricks of bright green, and the intensely blue flowers are carried on six-inch stems. The leaves of Mawson's Blue seem to last all the year and the flowers do not appear quite so spectacular with such a background of rough dark green.

Pulmonarias seed quite generously and the clumps can be divided as they grow bigger. Though these plants look right anywhere in a cottage garden they need more careful placing in a larger sphere. They are perfect in a wild or woodland garden, for their leaves get bigger as the season advances and compete successfully with the toughest grass. And for that reason they do not merit a front line position in the flower garden, where their coarse foliage will displease later in the year. They are a real pleasure in the spring garden and if they can be planted at the back of the border, against a wall or hedge, their bright little faces can be enjoyed with the knowledge that when the tiny leaves of babyhood grow coarse with middle-age they will blend and become absorbed with the summer growth of the other plants in the border.

MARGERY FISH, 1961

I have in my hand a small red poppy which I gathered on Whit Sunday on the palace of the Caesars. It is an intensely simple, intensely floral, flower. All silk and flame: a scarlet cup, perfect-edged all round, seen among the wild grass far away, like a burning coal fallen from heaven's altars. You cannot have a more complete, a more stainless, type of flower absolute; inside and outside, *all* flower. No sparing of colour anywhere – no outside coarsenesses – no interior secrecies; open as the sunshine that creates it; fine-finished on both sides, down to the extremest point of insertion on its narrow stalk; and robed in the purple of the Caesars ... We usually think of a poppy as a coarse flower; but it is the most transparent and delicate of all the blossoms of the field. The rest – nearly all of them – depend on the *texture* of their surfaces for colour. But the poppy is painted *glass*; it never glows so brightly as when the sun shines through it. Wherever it is seen – against the light or with the light – always, it is a flame, and warms the wind like a blown ruby.

JOHN RUSKIN, 1875–86

I should like to be strong-minded enough to dislike Dahlias and to shut the garden gate on one and all of them as a punishment to them and the raisers who have produced some of the horrors of modern garden nightmares. But as it would not make much stir in the world of Dahlias however tightly I barred it, and as I have a great affection for single Dahlias and all the true species I have been able to get, the family is still admitted. I cannot believe I shall ever be converted to a taste for Collarette Dahlias. The crumpled little pieces of flower that form the frill cause an itching in my fingers to pick them all out to see what the flower would look like without them; and latterly they have taken to themselves such appallingly virulent eye-jarring combinations of colour that I long to burn them, root and all. A screeching magenta Dahlia with a collarette of lemon yellow, or of a mixture of cerise and yellow, I think only fit to be grown in the garden of an asylum for the blind. Even the name annoys me; it suggests a sham affair, a dickey or some lace abomination, a middle-class invention to transform useful work-a-day clothes into a semblance of those of afternoon leisure.

Then we have a new race of Pæony-flowered Dahlias that get larger and more violently glaring each season, and their poor stalks seem more and more unable to hold up the huge targets that grow out of them. They must be matchless for saving labour in decorating for Harvest Thanksgivings, one pumpkin with two Pæony Dahlias and one bud would be enough for the pulpit, and an extra large one would be all-sufficient for the font. I saw some magnificently grown last season, but so huge that I could not resist asking, in as innocent a tone as I could manage, whether they fried or stewed them, and if the latter, were they best with a white sauce or brown gravy.

E. A. BOWLES, 1915

No old-time or modern garden is to me fully furnished without peonies. I would grow them in some corner of the garden for their splendid healthy foliage if they hadn't a blossom. The *Paeonia tenuifolia* in particular has exquisite feathery foliage. The great tree peony, which came from China, grows eight feet or more in height, and is a triumph of the flower world; but it was not known to the oldest front yards. Some of the tree peonies have finely displayed leafage of a curious and very gratifying tint of green ... The single peonies of recent years are of great beauty, and will soon be esteemed here as in China.

Not the least of the peony's charms is its exceeding trimness and cleanliness. The plants always look like a well-dressed, well-shod, well-gloved girl of birth, breeding, and of equal good taste and good health; a girl who can swim, and skate, and ride, and play golf. Every inch has a well-set, neat, cared-for look which the shape and growth of the plant keeps from seeming artificial or finicky.

No flower can be set in our garden of more distinct antiquity than the peony; the Greeks believed it to be of divine origin. A green arbor of the fourteenth century in England is described as set around with gillyflower, tansy, gromwell, and 'Pyonys powdered ay betwene' – just as I like to see peonies set to this day, 'powdered' everywhere between all the other flowers of the border.

ALICE MORSE EARLE, 1901

On the banks of the Bosphorus the tulip is the emblem of inconstancy; but it is also the symbol of the most violent love. The wild tulip is found in the fields of Byzantium, with its crimson petals and golden heart. The petals are compared to fire, and the yellow heart to brimstone; and, when presented by an admiring swain to his mistress, it is supposed to declare that such is the effect of the fair one's beauty that, if he sees her only for a moment, his face will be as fire, and his heart will be reduced to a coal.

The tulip was called *tulipan*, or *turban*, from the similarity of its corolla to the superb head-dress of the barbarous Turks, who almost worshipped its elegant stem and the beautiful vase-like flower which surmounts it. They never ceased to admire the gorgeous hues of gold and silver, of purple, lilac, and violet, of deep crimson and delicate rose-colour, with every possible variety of tint, which are harmoniously blended together, and spread over the rich petals of this splendid member of the court of Flora. The resemblance its shape bears to the turban is thus alluded to in Lalla Rookh: –

> What triumph crowds the rich divan to-day,
> With turbaned heads of every hue and race,
> Bowing before that veiled and awful face,
> Like tulip beds, of different shape and dyes,
> Bending beneath the invisible west wind's sighs.

Formerly a feast of tulips was celebrated in the seraglio of the Grand Seignior. Long galleries were erected, with raised seats, covered with the richest tapestry, presenting the appearance of an amphitheatre. On these were placed an almost infinite number of crystal vases, filled with the most beautiful tulips the world produced. In the evening the scene was splendidly illuminated; the wax tapers, as they gave light, emitted the most exquisite odours. To these were added lamps of the most brilliant colours, forming on all sides garlands of opal, emeralds, sapphires, diamonds, and rubies. Innumerable singing-birds, in cages of gold, roused by the splendour of the scene, mingled their warbling notes with the melodious harmony of instruments whose chords were tuned by invisible musicians. Showers of rose-water refreshed the air; and suddenly the doors were opened, and a number of young odalisks entered to blend the brilliancy of their charms and appearance with that of the enchanted scene.

In the centre of the seraglio a splendid pavilion shaded the Grand Seignior, who negligently reclined on costly skins; while the lords of his court, habited in their richest attire, were seated at his feet, to behold the dances of the lovely women of the court in all the luxurious display of their light and dazzling dresses. These sometimes encircled, and at others glided round, the vases of tulips whose beauty they sung. It was not seldom that a cloud rested on the sultan's brow; then he looked upon all around with a stern and severe aspect. What! could chagrin then enter the soul of that all-powerful mortal? Had he lost one of his provinces? Did he fear the revolt of his fierce janissaries? Ah no! two poor slaves alone had troubled his heart. He had observed, during the gaieties of the feast, a young page presenting a tulip to a beautiful girl who had captivated him. The sultan was ignorant of their secrets, but a vague feeling of inquietude took possession of his heart – jealousy tormented and beset him. But what is the jealousy of a sultan, or what are bolts and bars, against love? A look and a flower are enough for that wicked god to change a horrid seraglio into a place of delight, and to avenge beauty outraged by chains.

ROBERT TYAS, 1842

The greatest double Marigold hath many large, fat, broad leaves, springing immediatly from a fibrous or threddy root; the upper sides of the leaves are of a deepe greene, and the lower side of a more light and shining greene: among which rise up stalkes somewhat hairie, and also somewhat joynted, and full of a spungeous pith. The floures in the top are beautifull, round, very large and double, something sweet, with a certaine strong smell, of a light saffron colour, or like pure gold: from the which follow a number of long crooked seeds, especially the outmost, or those that stand about the edges of the floure; which being sowne commonly bring forth single floures, whereas contrariwise those seeds in the middle are lesser, and for the most part bring forth such floures as that was from whence it was taken.

This fruitfull or much bearing Marigold is likewise called of the vulgar sort of women, Jacke-an-apes on horse backe: it hath leaves, stalkes, and roots like the common sort of Marigold, differing in the shape of his floures, for this plant doth bring forth at the top of the stalke one floure like the other Marigolds; from the which start forth sundry other small floures, yellow likewise, and of the same fashion as the first, which if I be not deceived commeth to passe *per accidens*, or by chance, as Nature often-times liketh to play with other floures, or as children are borne with two thumbes on one hand, and such like, which living to be men, do get children like unto others; even so is the seed of this Marigold, which if it be sowen, it brings forth not one floure in a thousand like the plant from whence it was taken.

The Marigold floureth from Aprill or May even untill Winter, and in Winter also, if it bee warme.

The Marigold is called *Calendula*: it is to be seene in floure in the Calends almost of every moneth: it is also called *Chrysanthemum*, of his golden colour.

Columella in his tenth booke of Gardens hath these words;

> Stock-Gillofloures exceeding white,
> And Marigolds most yellow bright.

The floures and leaves of Marigolds being distilled, and the water dropped into red and watery eies, ceaseth the inflammation, and taketh away the paine.

Conserve made of the floures and sugar taken in the morning fasting, cureth the trembling of the heart.

The yellow leaves of the floures are dried and kept throughout Dutchland against Winter, to put into broths, in Physicall potions, and for divers other purposes, in such quantity, that in some Grocers or Spice-sellers houses are to be found barrels filled with them, and retailed by the penny more or lesse, insomuch that no broths are well made without dried Marigolds.

The common Africane, or as they vulgarly terme it French Marigold, hath small weake and tender branches trailing upon the ground, reeling and leaning this way and that way, beset with leaves consisting of many particular leaves, indented about the edges, which being held up against the sunne, or to the light, are seene to be full of holes like a sieve, even as those of Saint Johns woort: the floures stand at the top of the springy branches forth of long cups or husks, consisting of eight or ten small leaves, yellow underneath, on the upper side of a deeper yellow tending to the colour of a darke crimson velvet, as also soft in handling: but to describe the colour in words, it is not possible, but this way; lay upon paper with a pensill a yellow colour called

Masticot, which being dry, lay the same over with a little saffron steeped in water or wine, which setteth forth most lively the colour. The whole plant is of a most ranke and unwholesome smell, and perisheth at the first frost.

The unpleasant smel, especially of that common sort with single floures doth shew that it is of a poisonsome and cooling qualitie; and also the same is manifested by divers experiments: for I remember, saith *Dodonæus*, that I did see a boy whose lippes and mouth when hee began to chew the floures did swell extreamely; as it hath often happened unto them, that playing or piping with quils or kexes of Hemlockes, do hold them a while betweene their lippes: likewise he saith, we gave to a cat the floures with their cups, tempered with fresh cheese, shee forthwith mightily swelled, and a little while after died: also mice that have eaten of the seed thereof have been found dead. All which things doe declare that this herbe is of a venomous and poysonsome facultie; and that they are not to be hearkened unto, that suppose this herb to be a harmlesse plant: so to conclude, these plants are most venomous and full of poison, and therefore not to be touched or smelled unto, much lesse used in meat or medicine.

JOHN GERARD, 1597

Cowslips: Double Cowslips are fit to be planted in a Garden of pleasure for the use of their Flowers in sallets [salads], for the bedecking of the Garden, because they flower early when other Flowers are scarce, being once planted there they continue always; they never bear seed, therefore they must be planted.

The time of planting of them is either in the Spring or the Fall, the place is in the edge of the upper part of your borders, having prepared your ground then slip your plants into as many slips as you can, cutting off the top leaves within three inches of the root, and strain your line and prick them in three inches one from another where they will grow very well, if you water them, this must be while they are well rooted, and afterward they need no care but weeding, now remember to clip off the dead leaves and stocks after your Cowslips have done flowering, then the leaves will spring green and fresh again which is very pleasant to behold.

Daysies: There are three or four kinds, as the wild Daysie, the French Daysie, and the Garden double Daysie; the Garden Daysie it is I intend to treat of, of these there are two or three sorts of colours, but one in nature; the colours are these, the white, the red, the purple and the speckled: This flower never beareth any seed; the time of flowering is in May and June, a fine ornament to a Garden, and the flowers are used in Nosegayes: The branches of this flower dieth every year, and the root sendeth up young again; so where they are once planted they alwayes continue.

The place, time and manner of planting them, is as I told you of the Couslips, onely the choisest sorts be set in knots or beds, so I need not trouble my self nor you to give any farther reason for the ordering of them.

STEPHEN BLAKE, 1664

The last day of the month [January] sees our own real native snowdrop fully out, the foreign Elwesii having been in flower for some days.

Thick clumps of them are everywhere, and though we have quite a satisfactory amount of them in the garden here, their headquarters are at the Great House. Here they grow in their thousands, and in their thousands are sent to market, where we hope they cheer up the town-dwellers by telling them that spring is coming, and that the short dark days will soon be over. Tight little bunches of them, each surrounded by ivy leaves, are now on their way to Covent Garden, and, though they will not, like most country cousins, be taken to see *all* the sights, they will probably get a good glimpse of London life. They will look out from florists' windows on the hurrying crowds of Regent Street, and will see, from flower-sellers' baskets, the bewildering traffic of Piccadilly and Oxford Circus, and may even make the acquaintance of a Belisha beacon, which, in their ignorance, they will probably take for an overgrown and rather curious coloured globe artichoke. They will penetrate into sick rooms and hospitals, and adorn dining-room tables and drawing-room mantlepieces before they wither away and are collected by the dust cart.

But because the snowdrops, at the Great House, are picked and marketed, their beauty, as a whole, is somewhat spoiled, and to see them in perfection, though in lesser quantities, one must go to the churchyard, which they seem to have made their own, and where they seed about at will. They have massed themselves along the low wall that runs by the stream, and from here they send out scouts and establish outposts, which appear in twos and threes on the different graves. If by the following year these twos and threes are umolested, they turn themselves into little clumps.

The rabbits, which destroy everything else, leave the snowdrops alone, and we fear their only enemy is man, or, more probably, woman. For even these little snowdrops, beautifying God's acre, are sometimes considered to be common property, and are rooted up and carried off.

ETHEL ARMITAGE, 1936

The Crown Imperials are in full flower. They, like many other bulbs in this light soil, reproduce themselves so quickly that they want to be constantly lifted, the small bulbs taken away and put in a nursery (if you wish to increase your stock), and the large ones replaced, in a good bed of manure, where you want them to flower the following year. It is best, if possible, to do this in June, when the leaves have died down, but not quite disappeared so that the place is lost; one can, however, always find them in the autumn by their strong smell when the earth is moved beside them.

The orange Crown Imperials do best here, so, of course, I feel proudest of the pale yellow. Both colours are unusually good this year. In my youth they were rather sniffed at and called a cottage plant. I wonder if anyone who thought them vulgar ever took the trouble to pick off one of the down-hanging bells and turn it up to see the six drops of clear water in the six white cups with black rims? I know nothing prettier or more curious amongst flowers than this. I have not got the white one, but must try and get it; I am told it is very pretty, and so it must be, I should think.

MRS C. W. EARLE, 1897

I think I never saw a finer show of white Arums than we have just now. There is the grandest luxuriance of foliage, with thick tall stems, crowned by spathes in spiral lines of perfect grace. The rich texture of these flowers is marvellous; white as the drifted snow, with a lemon scent. Our success is perhaps due, not only to good management, but to what one may call, imported bulbs. Four years ago they were thrown out of a garden at Cannes, as worthless rubbish, on to the roadside. I passed that way one day, while a little peasant girl was collecting some of these bulbs in her pinafore. I asked her what they were? '*Des lis!*' she said. So I immediately gathered up some for myself, and they were done up in newspapers and packed in our trunks and brought home. In grim contrast to these joyous flowers of light is the Serpent Flower, a tropical member of the Arum family. I saw it, once only, eleven years ago, in the beautiful garden of Palazzo Orenga, at Mortola, near Ventimiglia. It grew on the edge of a ravine, under the deep shade of a low stone wall. Right up from a cluster of black-spotted leaves the centre spiral rose to about ten or twelve inches, bending over at the top into a sort of hood, like the hooded head of a cobra. The creature – flower, I cannot say – took the attitude exactly of a snake preparing to spring – the body marked and spotted the same as a snake, with the hood greyish-brown. The whole thing seemed something more than a good imitation only of the reptile whose name it bears. The first glance gave a sort of shock, as if on a sudden one had become aware of the actual presence at one's feet of a deadly serpent; and yet this terrifying object is, I believe, used by the Indians as an antidote to snake-bite.

MRS E. V. BOYLE, 1884

The rose bushes were planted along the sides of the road which ran through our village and were greatly admired by the passersby, but it was strongly impressed upon us that a rose was useful, not orna-mental. It was not intended to please us by its color or its odor, its mission was to be made into rosewater, and if we thought of it in any other way we were making an idol of it and thereby imperiling our souls. In order that we might not be tempted to fasten a rose upon our dress or to put it into water to keep, the rule was that the flower should be plucked with no stem at all. We had only crimson roses, as they were supposed to make stronger rosewater than the paler varieties. This rosewater was sold, of course, and was used in the community to flavor apple pies. It was also kept in store at the infirmary, and although in those days no sick person was allowed to have a fresh flower to cheer him, he was welcome to a liberal supply of rosewater to bathe his aching head.

A SISTER OF THE SHAKER COMMUNITY, 1906

My early recollections of church-going are associated with roses. We went every Sunday morning to an old Queen Anne church: ours was a square pew; the pew-opener, a woman, walked before us, opened the pew door and shut us in. We sat round facing one another, but could not see anything except the gallery having the royal arms in the centre, the children who sat there with the village schoolmaster, who was also parish clerk and gave out the hymns. When standing-up time came I had to stand on the seat to see over the top of the pew. In a neighbouring pew there was a gentleman who appeared every Sunday with a rose in his buttonhole; I admired that rose, and resolved to wear as good if not a better one the next Sunday. During the week I was on the look-out for a suitable one, and when Sunday came again it was gathered – Moss, White-crested Moss, Red Provence at first, and then Baron de Maynard or Boule de Neige were favourites. I appeared with my bloom, and when the time came to mount the seat compared it with the rose in the buttonhole of my rival. The result of the judging was usually adverse to me, but I always went home hoping for better luck next time. My flowers were handicapped by the staging; you see, I was in petticoats at first and wore a light-coloured Norfolk jacket, large mother-of-pearl buttons down the front, and a belt. My rival had a black coat, and the rose had a buttonhole all to itself; there is nothing like black to set off a rose, especially when added to this the flower did not have to share the buttonhole with a large button. I was quite aware of the drawback, and longed for the time when I might have a cloth jacket with a buttonhole at the side.

Then came a time when I had a small garden of my own. There were three standards in it – red roses; I did my own pruning, but they hardly ever gave me a Sunday buttonhole; the situation was too shady, and there were laurels close by. When about twelve years of age my father showed me how to bud. My first attempt was to bud a White-crested Moss; it was a failure, however; father said the stock did not run, but I am now inclined to think the budder bruised the bud.

My father used to go to the Crystal Palace rose show, and some-times took me with him. On one occasion I had in my buttonhole a grand bloom of Marie Baumann. I remember that flower so well, and how the people seemed to look at it as we walked over London Bridge together; I felt sure by the way they looked that we should not see a better one at the show. I liked those visits; they meant more standards in the autumn from Rivers. Not that father relied altogether on what he had seen at the show; before ordering them he would consult a man who budded for Rivers. He kept pace with the times, and had a book which he read a good deal, 'The Rose Amateur's Guide', by Thomas Rivers: I have it still. Some of the roses we then had as standards were, Senateur Vaisse, Générale Jacqueminot, John Hopper, Comtesse de Chabrillant, and Madame Bravy.

When going back to school in September, I used to take with me a pointed flower of Souvenir de la Malmaison, packed in an empty barley-sugar tin; it kept fresh in that tin box for a long time, and daily I would take it out, admire it, and recall happy memories of home of which it reminded me. Brown and battered I brought it home with me at Christmas. The peculiar perfume of a Souvenir de la Malmaison – a kind of beery smell – reminds me to this day of the rose in the barley-sugar box; and although so long ago the standard from which those blooms were gathered is with us still.

My father, however, never exhibited, but in the summer of 1874, the summer that followed his death, I ventured on my first attempt. I went round the standards the day before the show and found we could just get twelve varieties. Assisted by the gardener we borrowed a chrysanthemum stand with legs and tubes; we covered the board with lycopodium on which the roses rested, cut with foot-stalks only, no foliage; the man said that was the proper way to do it, but having been to the Crystal Palace, I had my doubts. Although we had only the twelve roses and did not take any extra blooms, yet the stand won second prize. That did it; from henceforth I was on the warpath; fifty standards were ordered from Rivers, and a piece of the kitchen garden was prepared where they could grow all by themselves free from gooseberry bushes. The next year I went to two local shows, and in 1876 to the Alexandra Palace and Crystal Palace exhibitions, where Mr Benjamin R. Cant and Mr George Prince took me in hand, giving me advice in staging and kindly encouragement. So that now I can look back upon thirty-two years of rose-showing, and to having exhibited at every metropolitan show of the National Rose Society, including the first, held at St James's Hall, and at which I was awarded second prize for twelve roses in a class in which there were forty competitors.

REV. JOSEPH PEMBERTON, 1908

THE SICK ROSE

O rose, thou art sick!
The invisible worm
That flies in the night,
In the howling storm,

Has found out thy bed
Of crimson joy,
And his dark secret love
Does thy life destroy.

WILLIAM BLAKE, 1794

The most powerful and far-reaching scent in the garden by daylight is given off by *Humea elegans*, not only by the flowers – plumes of chestnut-coloured Pampas grass – but also by the stems and leaves. It resembles incense so closely that it once caused a country vicar to be suspected of ritualistic practices; and one day, as I was walking down Victoria Street, some hours after handling a leaf of it at Vincent Square, a friend who was with me remarked, 'What a long way the smell of incense carries from Westminster Cathedral!'

JASON HILL, 1932

Among the many wonders of the vegetable world are the flowers that hang their heads and seem to sleep in the daytime, and that awaken as the sun goes down, and live their waking life at night. And those that are most familiar in our gardens have powerful perfumes, except the Evening Primrose (*Œnothera*), which has only a milder sweetness. It is vain to try and smell the night-given scent in the daytime; it is either withheld altogether, or some other smell, quite different, and not always pleasant, is there instead. I have tried hard in daytime to get a whiff of the night sweetness of *Nicotiana affinis*, but can only get hold of something that smells like a horse! Some of the best of the night-scents are those given by the Stocks and Rockets. They are sweet in the hand in the daytime, but the best of the sweet scent seems to be like a thin film on the surface. It does not do to smell them too vigorously, for, especially in Stocks and Wallflowers, there is a strong, rank, cabbage-like under-smell. But in the sweetness given off so freely in the summer evening there is none of this; then they only give their very best.

But of all the family, the finest fragrance comes from the small annual Night-scented Stock (*Matthiola bicornis*), a plant that in daytime is almost ugly; for the leaves are of a dull-grey colour, and the flowers are small and also dull-coloured, and they are closed and droop and look unhappy. But when the sun has set the modest little plant seems to come to life; the grey foliage is almost beautiful in its harmonious relation to the half-light; the flowers stand up and expand, and in the early twilight show tender colouring of faint pink and lilac, and pour out upon the still night-air a lavish gift of sweetest fragrance; and the modest little plant that in strong sunlight looked unworthy of a place in the garden, now rises to its appointed rank and reigns supreme as its prime delight.

GERTRUDE JEKYLL, 1899

Who loves a garden, loves a greenhouse too.
Unconscious of a less propitious clime,
There blooms exotic beauty, warm and snug,
While the winds whistle and the snows descend.
The spiry myrtle with unwithering leaf
Shines there and flourishes. The golden boast

Of Portugal and western India there,
The ruddier orange and the paler lime,
Peep through their polished foliage at the storm,
And seem to smile at what they need not fear.
The amomum there with intermingling flowers
And cherries hangs her twigs. Geranium boasts
Her crimson honours, and the spangled beau,
Ficoides, glitters bright the winter long.
All plants, of every leaf that can endure
The winter's frown, if screened from its shrewd bite,
Live there and prosper. Those Ausonia claims,
Levantine regions these; the Azores send
Their jessamine, her jessamine remote
Caffraria: foreigners from many lands,
They form one social shade, as if convened
By magic summons of the Orphean lyre.
Yet just arrangement, rarely brought to pass
But by a master's hand, disposing well
The gay diversities of leaf and flower,
Must lend its aid to illustrate all their charms,
And dress the regular yet various scene.
Plant behind plant aspiring, in the van
The dwarfish, in the rear retired, but still
Sublime above the rest, the statelier stand.

WILLIAM COWPER, 1785

It is a greater act of faith to plant a bulb than to plant a tree.

CLARE LEIGHTON, 1935

As soon as the rain holds up for a day or two and the soil dries under the sun I want to set about the adorable misery of planting my bulbs. I get tired, bored, sick to death of the job, and every year I find afresh how worth while it is – for the pictures come flashing to and fro under the trowel and aching muscles – in gold and ivory, purple and red. Imagination flowers them; and then all the riches and the wonder of the world are mine for a moment. In the Spring the dreams come true. All the toil is worth while.

MARION CRAN, 1929

The filling of so many square yards of prepared soil with so many thousands of expensive bulbs, to yield a certain shade of colour for a fortnight and then to be pulled up to make place for another massing, gives me a sort of gardening bilious attack, and a feeling of pity for the plants and contempt for the gardening skill that relies upon Bank of England notes for manure.

E. A. BOWLES, 1914

Look ... at the wide and bare belts of grass that wind in and around the shrubberies in nearly every country place; frequently, they never display a particle of plant-beauty, and are merely places to be roughly mown now and then. But if planted here and there with the Snowdrop, the blue Anemone, Crocus, Scilla, and Winter Aconite, they would in spring surpass in charms the gayest of 'spring gardens'. Cushioned among the grass, the flowers would unfold prettier than they can in the regulation sticky earth of a border; in the grass of spring, their natural bed, they would look far better than they ever do on the brown earth of a garden. Once carefully planted, they – while an annual source of the greatest interest – occasion no trouble whatever.

Their leaves die down so early in spring that they would not interfere with the mowing of the grass, and we should not attempt to mow the grass in such places till the season of vernal flowers had passed. Surely it is enough to have a portion of lawn as smooth as a carpet at all times, without shaving off the 'long and pleasant grass' of the other parts of the grounds. It would indeed be worth while to leave many parts of the grass unmown for the sake of growing many beautiful plants in it. If in a spot where a wide carpet of grass spreads out in the sheltered bay of a plantation, there be dotted the blue Apennine Anemone, any Snowdrops, the Snowflake, Crocuses in variety, Scillas, Grape Hyacinths, many Narcissi, the Wood Anemone, and any other Spring flowers liking the soil, we should have a picture of vernal beauty, the flowers relieved by grass, and the whole devoid of man's weakness for tracing wallpaper patterns where everything should be varied and changeful.

WILLIAM ROBINSON, 1870

CORNISH FLOWER-FARM

Here where the cliff rises so high
The sea below fills half the sky
And ships hang in mid-air,
Set on the cliff-face, square by square,
Walls of veronica enclose
White gladioli in their neat rows
And blue and golden irises;
But though the walls grow tall as trees,
Some flowers from their quiet quillets pass
To mix with wayside weeds and grass,
Like nuns that from their strict retreats
Go visiting the poor in their plain streets.

ANDREW YOUNG, 1950 (1981)

*K*napp *Hill Nursery; Mr Waterer. August 6.* – We had heard much in London, and from various gardeners in the country, of the splendid collection of new seedling azaleas which flowered here in June last, not one of which is yet given out to the trade; but, of course, at this season we could only see the foliage. Among other things we noted Andromeda arborea, 10ft high, and finely in flower; Vaccinium Arctostaphylos, the Madeira bilberry, 6ft high, and richly covered with fruit; another species, unknown, bearing very large fruit. Both species well deserve culture, where peat earth is not scarce, as fruits for tarts and for eating with cream, like other bilberries. Magnolia auriculata, very luxuriant; measured one of the leaves, and found it 22in long, and

11in wide. Lilium superbum, 10ft high, coming into flower. The great art in getting this species to flower well, as Mr Cameron of Bury Hill informed us, is to keep the bulbs single, by taking them up, separating, and replanting. It is evident that, by this practice, the greatest possible supply of nourishment will be obtained by each plant. Phlox Thomsoni, a new variety, in flower. Daphnia collina, a variety with striped leaves. This nursery excells in the management of hedges, which are in some cases 8 or 10ft high, and not more than 8 or 10in thick: but, in general, it is not quite so neat and orderly as we could wish; and though we have never seen the weeds exceed the economic point, we would rather see weeding carried lower. We never yet knew a nursery or market-garden, where any money was made, that was not kept *orderly*, at all events, and most of them even *neatly*. We do not say that much is wanting at Knapp Hill; but still we should like to see both principles pushed farther; a good many of the old things grubbed up, the walks and compartments more correctly lined out, and no weeds ever suffered to grow above an inch high. We hint this with the more confidence, knowing that Mr Waterer will take it in good part, and that it will be in his favour with the hundreds of gardeners and gentlemen that will come from all quarters next June to see the bloom of new azaleas.

JOHN CLAUDIUS LOUDON, 1829

Sir, – I Rec'd your Letter & your Baskett of flowers per Capt. Morris, & have Desired Francis Wilks Esqu. to pay you £26 for them *Though they are Every one Dead*. The Trees I Rec'd Last Year are above half Dead too, – the Hollys all Dead but one, & worse than all is the Garden Seeds and Flower Seeds which you Sold Mr Wilks for me an Charged me £6. 4s. 2d. Sterling were not worth one farthing. Not one of all the Seeds Came up Except the Asparrow Grass, So that my Garden is Lost for me this Year. I Tryed the Seeds both in Town and Country & all proved alike bad. I spared Mr Hubbard part of them *and they All Served him the Same.* I think Sir you have not done well by me in this thing, for me to send 1,000 leagues and Lay out my money & be so used & Disapointed is very hard to Bare, & so no doubt but you will Consider the matter & Send me over Some more of the Same Sort of Seeds that are Good & Charge me nothing for them, – if you don't I shall think you have imposed upon me very much, & t'will Discourage me from Sending again for Trees or Seeds from you. I Conclude

Your Humble Serv't

T. H.

P.S. *The Tulip Roots you were pleased to make a present off to me are all Dead as well.*

THOMAS HANCOCK, 1737

S eeds of all Sorts (and consequently the Sellers thereof) are subject to so many Calamities and Inconveniences, both in the saving, selling, buying, and sowing, (some accidental and some natural) that my Reader will not, I humbly hope, be displeased if I set this Affair, as well as I can, in the most genuine Light, with such Directions as shall be judged most proper for the avoiding these misfortunes, as far at least as human Foresight can prevent; for whatever some Persons may think it cannot but be a matter of great Concern to every honest Man that is concern'd in the Seed or Nursery Trade, to find that by any Means (even tho' it be not his own fault) either his Seeds or Trees miscarry, or prove naught. And in this Respect, if a Man means well, he may justly be engaged in one of the most unhappy, tho' most useful

Scenes of Life, and the Reasons which I give for it are these which follow.

First, The Badness of some Years is such, that many sorts of Seeds do not ripen well, and then the Seeds in the top Husks, (as in many Grass-Seeds as well as those growing in the Garden) are small, and not at their full Perfection, and then the Seedsman is immediately blam'd, and perhaps stigmatiz'd with such names as he least deserves.

A second Reason is, that those whose Business it is to save Seeds, are not so careful as they ought; and if they are in their own Nature hot, such as Onions, and the like, if put up into too thick Heaps, they heat and the Vegetative Property soon spoils, tho' at the same time the Seed will look bright and well, and the Seedsman comes in for his share of Reproach, tho' this evil cannot be foreseen till too late.

To the two Reasons before-going, may well be added a third, and that is, some Seeds which are imported from abroad, either by the length of their Passage, or the Perfidiousness of those who sell them, are spoil'd and good for nothing; and this often happens to private Gentlemen, who have them from abroad for their own use, as well as Traders; but even here again the Seller, tho' he never saw the saving of them, nor is it possible by the sight of them, to know the good Seed from the bad, either as to their goodness or their kinds, yet he is peremptorily blamed as a Knave; and tho' this is not common to Persons of Understanding or Judgment, yet it too often happens to the fair Trader who certainly if he has any Regard for himself, would wish it were otherwise. And this Complaint affects him in Seeds sav'd in his own Country, if he be not by whilst it is doing.

A fourth Reason for the Miscarriage of Crops, both in husbandry and Gardening, is the Negligence or Unskilfullness, of those to whose Management those first Rudiments are committed; for either their Ground (if it be wet and stiff) is not well fallow'd and trench'd, or else the Seed is sown at a wrong time, too soon or perhaps too late; and tho', added to all, it is true that the Spring Seasons of this Climate are so uncertain, that even the most sagacious understanding Person may err in some of these Extreams, on which Account many very honest and laborious Gardeners and Farmers, about the Neat Houses, and in the Fields at and about Vaux-Hall and Battersea, notwithstanding all their Care, are often obliged to crop their Grounds three or four times; yet when the like Case happens to some hot headed inconsiderate Person, then the Seedsman must come in for a very large share of his Noise and Nonsense, tho' not at all concern'd in the ill Effects thereof.

I wish there was not too much Reason to add a fifth Cause of the Unhappiness that the Seedsmen lie under, from the Unwillingness as well as ill conduct of some of these Husbandmen and Gardeners who are not in the Interest of those Seedsmen, and who indeed hate that their Masters should purchase their own Goods, because they think it an Intrusion into their Province, and perhaps may debar them of some Perquisites they propose to themselves. An Instance of this kind I remember happened the last Summer; when a servant in this way to a Gentleman, not imagining he should be detected in it, assured his Master that the Cauliflower Seed he had from such a one was not good, when upon a Perusal of the Account, that Person very luckily had not sold it the Gentleman, but was of the Gardener's own procuring. But it very often happens, that the good Seeds are either changed or spoil'd, that the Odium may be cast on the Seller.

It will be impossible, without great Circumspection and Attention, in any Master to remedy all these Inconveniencies, but some of them he may, by bringing his Ground into good Tillage, by Winter fallowing, trenching, &c. and because most Grounds which are stiffish, strong, and rich, are subject to Weeds, Worms, Grubs, Snails, and other Vermin, many of which are imperceptible without the help of a Microscope; it would be well for him to dress his Land with Lime, Soot, or any other Manure of a hot Quality, before he sows his Seeds, about some two or three Months, for that will destroy those infectious Animals that spoil the sprouting tender Seeds, and make the Ground mellow; and the Reason for the sowing of these things so long before is, that the Fire thereof may be so abated as that it may not burn up the tender Artichoaks, Asparagus, Beans, Pease, or Kitchen Seeds and if Fruit-Trees are planted or sown, then you may defer the laying any of these Manures upon them, 'till they are just coming up; but then that must be done with Judgment, for I have known some young unrooted Fruit-Trees burnt up with them, and those Helps, except it be to Vines, and other things which draw up vast Quantities of Nourishment, ought to be chiefly to old decaying Trees, and not those which are in Youth and Vigour.

If rainy Weather should not immediately follow after the sowing of small Seeds, especially in the Garden, the Use of the Water-Cart, or Water-Barrow and Pot, is to be recommended, otherwise all those Seeds which lie near the top of the Ground will sprout out by the mighty Dews, and be destroy'd by the drying parching Winds of the ensuing Day, tho' at that time not visible to common Observation; and here the Seller also lies under the lash.

To conclude, tho' I can't take upon me to justify all who follow the Employment I have been mentioning, or say that some of them do not knowingly and willfully sell those Commodities that they are sure will not grow; yet I can solemnly declare, that I never directly nor indirectly us'd that Practice, *i.e.* in the buying and selling that which I certainly knew was not good; on which Account I have lost the taking of several Sums of Money. Nor am I ever sanguine enough to vindicate the best things, on the contrary, wherever I have had the least occasion to distributing the Goodness of the Commodities, I have told Gentlemen, and have publish'd Advertisements of it, as soon as I have made discovery of it. And the Encouragement, which I have met with, even beyond Expectation, in my present Employment is such, that I can't finish this ... without returning Thanks for it, with a Promise that for the future it shall be my endeavour to deserve the favour of my kind Correspondents.

STEPHEN SWITZER, 1731

As I sat in the porch, I heard the voices of two or three persons, who seemed very earnest in discourse. My curiosity was raised when I heard the names of Alexander of the Great and Artaxerxes; and as their talk seemed to run on ancient heroes, I concluded there could not be any secret in it; for which reason I thought I might very fairly listen to what they said.

After several parallels between great men, which appeared to me altogether groundless and chimerical, I was surprised to hear one say, that he valued the Black Prince more than the Duke of Vendosme. How the Duke of Vendosme should become a rival of the Black Prince, I could not conceive: and was more startled when I heard a second affirm, with great vehemence, that if the Emperor of Germany was not going off, he should like him better than either of them. He added, that though the season was so changeable, the Duke of Marlborough was in blooming beauty. I was wondering to myself from whence they had received this odd intelligence: especially when I heard them mention the names of several other generals, as the Prince of Hesse and the King of Sweden, who, they said, were both running away. To which they added, what I entirely agreed with them in, that the Crown of France was very weak, but that the Marshal Villars still kept his colours. At last, one of them told the company, if they would go along with him, he would show them a Chimney-Sweeper and a Painted Lady in the same bed, which he was sure would very much please them. The shower, which had driven them as well as myself into the house, was now over; and as they were passing by me into the garden, I asked them to let me be one of their company.

The gentleman of the house told me, 'if I delighted in flowers, it would be worth my while; for that he believed he could show me such a blow of tulips as was not to be matched in the whole country'.

I accepted the offer, and immediately found that they had been talking in terms of gardening, and that the kings and generals they had mentioned were only so many tulips, to which the gardeners, according to their usual custom, had given such high titles and appellations of honour . . .

I accidentally praised a tulip as one of the finest I ever saw; upon which they told me it was a common Fool's Coat. Upon that I praised a second, which it seems was but another kind of Fool's Coat . . . The gentleman smiled at my ignorance. He seemed a very plain honest man, and a person of good sense, had not his head been touched with that distemper which Hippocrates calls the Tulippomania; insomuch that he would talk very rationally on any subject in the world but a tulip.

He told me, 'that he valued the bed of flowers that lay before us, and was not above twenty yards in length and two in breadth, more than he would the best hundred acres of land in England', and added, 'that it would have been worth twice the money it is, if a foolish cook-maid of his had not almost ruined him the last winter, by mistaking a handful of tulip roots for a heap of onions, and by that means,' says he, 'made me a dish of pottage that cost me above a thousand pounds sterling'.

JOSEPH ADDISON, 1710

Opening out on the long piazza over the flower beds, and extending almost its whole length, runs the large, light, airy room where a group of happy people gather to pass the swiftly flying summers here at the Isles of Shoals. This room is made first for music; on the polished floor is no carpet to muffle sound, only a few rugs here and there, like patches of warm green moss on the pine-needle color given by the polish to the natural hue of the wood. There are no heavy draperies to muffle the windows, nothing to absorb the sound. The piano stands midway at one side; there are couches, sofas with pillows of many shades of dull, rich color, but mostly of warm shades of

green. There are low bookcases round the walls, the books screened by
short curtains of pleasant olive-green; the high walls to the ceiling are
covered with pictures, and flowers are everywhere. The shelves of the
tall mantel are splendid with massed nasturtiums like a blazing torch,
beginning with the palest yellow, almost white, and piled through every
deepening shade of gold, orange, scarlet, crimson, to the blackest red;
all along the tops of the low bookcases burn the fires of marigolds,
coreopsis, large flowers of the velvet single dahlias in yellow, flame,
and scarlet of many shades, masses of pure gold summer chrysanthemums, and many more – all here and there interspersed with
blossoming grasses for a touch of ethereal green. On one low bookcase are Shirley poppies in a roseate cloud. And here let me say that the
secret of keeping poppies in the house two whole days without fading
is this: they must be gathered early, before the dew has dried, in the
morning. I go forth between five and six o'clock to cut them while yet
their gray-green leaves are hoary with dew, taking a tall slender pitcher
or bottle of water with me into the garden, and as I cut each stem
dropping the flower at once into it, so that the stem is covered nearly
its whole length with water; and so on till the pitcher is full. Gathered
in this way, they have no opportunity to lose their freshness, indeed,
the exquisite creatures hardly know they have been gathered at all.
When I have all I need, I begin on the left end of this bookcase, which
most felicitously fronts the light, and into the glasses put the radiant
blossoms with an infinite enjoyment of the work. The glasses (thirty-
two in all) themselves are beautiful: nearly all are white, clear and pure,
with a few pale green and paler rose and delicate blue, one or two of
richer pink, all brilliantly clear and filled with absolutely colorless
water, through which the stems show their slender green lengths. Into
the glasses at this end on the left I put first the dazzling white single
poppy, the Bride, to lead the sweet procession – a marvelous blossom,
whose pure white is half transparent, with its central altar of ineffable
green and gold. A few of these first, then a dozen or more of delicate
tissue paper-like blossoms of snow in still another variety (with petals
so thin that a bright color behind them shows through their filmy
texture); then the double kind called Snowdrift, which being double

makes a deeper body of whiteness flecked with softest shadow. Then I begin with the palest rose tints, placing them next, and slightly mingling a few with the last white ones – a rose tint delicate as the palm of a baby's hand; then the next, with a faint suffusion of a blush, and go on to the next shade, still very delicate, not deeper than the soft hue on the lips of the great whelk shells in southern seas; then the damask rose color and all tints of tender pink, then the deeper tones to clear, rich cherry, and on to glowing crimson, through a mass of this to burning maroon.

The flowers are of all heights (the stems of different lengths), and, though massed, are in broken and irregular ranks, the tallest standing a little over two feet high. But there is no crushing or crowding. Each individual has room to display its full perfection. The color gathers, softly flushing from the snow white at one end, through all rose, pink, cherry, and crimson shades, to the note of darkest red; the long stems of tender green showing through the clear glass, the radiant tempered gold of each flower illuminating the whole. Here and there a few leaves, stalks, and buds (if I can bring my mind to the cutting of these last) are sparingly interspersed at the back. The effect of this arrangement is perfectly beautiful. It is simply indescribable, and I have seen people stand before it mute with delight. It is like the rose of dawn.

To the left of this altar of flowers is a little table, upon which a picture stands and leans against the wall at the back. In the picture two tea roses long since faded live yet in their exquisite hues, never indeed to die. Before this I keep always a few of the fairest flowers, and call this table the shrine. Sometimes it is a spray of Madonna lilies in a long white vase of ground glass, or beneath the picture in a jar of yellow glass floats a saffron-tinted water lily, the chromatella, or a tall sapphire glass holds deep blue larkspurs of the same shade, or in a red Bohemian glass vase are a few carmine sweet peas, another harmony of color, or a charming dull red Japanese jar holds a few nasturtiums that exactly repeat its hues. The lovely combinations and contrasts of flowers and vases are simply endless.

On another small table below the 'altar' are pink water lilies in pink glasses and white ones in white glasses; a low basket of amber glass is filled with the pale turquoise of forget-me-nots, the glass is iridescent and gleams with changing reflections, taking tints from every color near it. Sweet peas are everywhere about and fill the air with fragrance; orange and yellow Iceland poppies are in tall vases of English glass of light green. There is a large, low bowl, celadon-tinted, and decorated with the boughs and fruit of the olive on the gray-green background. This is filled with magnificent Jacqueminot roses, so large, so deep in color as to fully merit the word. Sometimes they are mixed with pink Gabrielle de Luizets and old-fashioned damask roses, and the bowl is set where the light falls just as it should to give the splendor of the flowers its full effect. In the center of a round table under one of the chandeliers is a flaring Venice glass as pure as a drop of dew and of a quaintly lovely shape; on the crystal water therein lies a single white water lily, fragrant snow and gold. By itself is a low vase shaped like a magnolia flower, with petals of light yellow deepening in color at the bottom, where its calyx of olive-green leaves clasps the flower. This has looking over its edge a few pale yellow nasturtiums of the Asa Gray variety, the lightest of all. With these, one or two of a richer yellow (Dunnett's Orange), the flowers repeating the tones of the vase, and with them harmoniously blending. A large pearly shell of the whelk tribe was given me years ago. I did not know what to do with it. I do not like flowers in shells as a rule, and I think the shells are best on the beach where they belong, but I was fond of the giver, so I sought some way of utilizing the gift. In itself it was beautiful, a mass of glimmering rainbows. I bored three holes in its edge and suspended it from one of the severely simple chandeliers with almost invisible wires. I keep it filled with water and in it arrange sometimes clusters of monthly honeysuckle sparingly; the hues of the flowers and the shell mingle and blend divinely. I get the same effect with hydrangea flowers, tints and tones all melt together; so also with the most delicate sweet peas, white, rose, and lilac; with these I take some lengths of the blossoming wild cucumber vine with its light clusters of white flowers, or the white clematis, the kind called 'Traveler's Joy', and weave it lightly about the shell, letting it creep over one side and, running up the wires, entirely conceal them; then it is like a heavenly apparition afloat in mid air. Sometimes the tender mauve and soft rose and delicate blues of the exquisite little Rose Campion, or Rose of Heaven, with its grassy foliage, swing in this rainbow shell, making another harmony of hues.

Sometimes it is draped with wild morning glory vines which are gathered with their buds at evening; their long wiry stems I coil in the water, and arrange the graceful lengths of leaves and buds carefully, letting a few droop over the edge and twine together beneath the shell, and some run up to the chandelier and conceal the wires. The long smooth buds, yellow-white like ivory, deepen to a touch of bright rose at the tips close folded. In the morning all the buds open into fair trumpets of sea-shell pink, turning to every point of the compass, an exquisite sight to see. By changing the water daily these vines last a week, fresh buds maturing and blossoming every morning.

Near my own seat in a sofa corner at one of the south windows stands yet another small table, covered with a snow-white linen cloth embroidered in silk as white and lustrous as silver. On this are gathered every day all the rarest and loveliest flowers as they blossom, that I may touch them, dwell on them, breathe their delightful fragrance and adore them. Here are kept the daintiest and most delicate of the vases which may best set off the flowers' loveliness – the smallest of the collection, for the table is only large enough to hold a few. There is one slender small tumbler of colorless glass, from the upper edge of which a crimson stain is diffused half way down its crystal length. In this I keep one glowing crimson Burgundy rose, or an opening Jacqueminot bud; the effect is as if the color of the rose ran down and dyed the glass crimson. It is so beautiful an effect one never wearies of it. There is a little jar of Venice glass, the kind which Browning describes in 'The Flight of the Duchess' –

> With long white threads distinct inside,
> Like the lake-flower's fibrous roots that dangle
> Loose such a length and never tangle.

This is charming with a few rich pinks of different shades. Another Venice glass is irregularly bottle-shaped, bluish white with cool sea-green reflections at the bottom, very delicate, like an aqua-marine. It is lightly sprinkled with gold dust throughout its whole length; toward the top the slender neck takes on a soft touch of pink which meets and mingles with the Bon Silene or La France rose I always keep in it. Another Venice glass still is a wonder of iridescent blues, lavenders, gray, and gold, all through, with a faint hint of elusive green. A spray of heaven-blue larkspur dashed with rose is delicious in this slender shape, with its marvelous tints melting into the blue and pink of the fairy flowers.

A little glass of crystal girdled with gold holds pale blue forget-me-nots; sometimes it is rich with orange and yellow erysimum flowers. In a tall Venetian vase of amber a *Lilium auratum* is superb. A low jar of opaque rose-pink, lost at the bottom in milky whiteness, is refreshing with an old-fashioned damask rose matching its color exactly. This is also exquisite with one pink water lily. The pink variety of the Rose Campion is enchanting in this low jar. A tall shaft of ruby glass is radiant with poppies of every shade of rose and lightest scarlet, with the silvery green of a few oats among them. A slender purple glass is fine with different shades of purple and lilac sweet peas, or one or two purple poppies, or an aster or two of just its color, but there is one long gold-speckled Bohemian glass of rich green which is simply perfect for any flower that blows, and perfect under any circumstances. A half dozen Iceland poppies, white, yellow, orange, in a little Japanese porcelain bottle, always stand on this beautiful table, the few flecks of color on the bottle repeating their tints. I never could tell half the lovely combinations that glow on this table all summer long.

CELIA THAXTER, 1894

CHAPTER FIVE

The Gardens

The Sunk Garden,
Bryan's Ground.
Simon Dorrell.

A garden is not made in a year; indeed it is never made in the sense of finality. It grows, and with the labour of love should go on growing.

FREDERICK EDEN, 1903

THE GLORY OF THE GARDEN

Our England is a garden that is full of stately views,
Of borders, beds and shrubberies and lawns and avenues,
With statues on the terraces and peacocks strutting by;
But the Glory of the Garden lies in more than meets the eye.

For where the old thick laurels grow, along the thin red wall,
You find the tool- and potting-sheds which are the heart of all;
The cold-frames and the hot-houses, the dungpits and the tanks,
The rollers, carts and drain-pipes, with the barrows and the planks.

And there you'll see the gardeners, the men and 'prentice boys
Told off to do as they are bid and do it without noise;
For, except when seeds are planted and we shout to scare the birds,
The Glory of the Garden it abideth not in words.

And some can pot begonias and some can bud a rose,
And some are hardly fit to trust with anything that grows;
But they can roll and trim the lawns and sift the sand and loam,
For the Glory of the Garden occupieth all who come.

Our England is a garden, and such gardens are not made
By singing: – 'Oh, how beautiful!' and sitting in the shade,
While better men than we go out and start their working lives
At grubbing weeds from gravel-paths with broken dinner-knives.

There's not a pair of legs so thin, there's not a head so thick,
There's not a hand so weak and white, nor yet a heart so sick,
But it can find some needful job that's crying to be done,
For the Glory of the Garden glorifieth every one.

Then seek your job with thankfulness and work till further orders,
If it's only netting strawberries or killing slugs on borders;
And when your back stops aching and your hands begin to harden,
You will find yourself a partner in the Glory of the Garden.

Oh, Adam was a gardener, and God who made him sees
That half a proper gardener's work is done upon his knees,
So when your work is finished, you can wash your hands and pray
For the Glory of the Garden, that it may not pass away!
And the Glory of the Garden it shall never pass away!

RUDYARD KIPLING, 1911

After what I have said of the Number and Beauty of the *Persian* Flowers, one would be very apt to imagine, that they must of course have the finest Gardens in the World; but it is no such thing; on the Contrary I have found it to be a general Rule, that where

Nature is most Easy and Fruitful, they are very raw and unskilful in the Art of Gardening. This comes to pass, by reason, that in those Places, where Nature plays so excellently the Part of a Gardner, if I may be permitted to use the Expression, Art hath in a Manner nothing to do. The Gardens of the *Persians*, commonly consist of one great Walk, which parts the Garden, and runs on in a straight Line, border'd on each side by a Row of Plantanes, with a Bason of Water in the middle of it, made proportionable in Bigness to the Garden, and likewise of two other little Side-Walks, the Space between them is confusedly set with Flowers, and planted with *Fruit-Trees*, and *Rose-Bushes*; and these are all the Decorations they have. They don't know what *Parterres* and *Green-Houses*, what *Wildernesses* and *Terraces*, and the other Ornaments of our Gardens are. The most particular Reason one can assign for this, is, that the *Persians* don't walk so much in Gardens as we do, but content themselves with a bare Prospect, and breathing the fresh Air: For this End, they set themselves down in some part of the Garden, at their first coming into it, and never move from their Seats till they are going out of it.

SIR JOHN CHARDIN, 1686

From *Lugane* I went to the *Lago Maggiore*, which is a great and noble Lake, it is six and fifty Miles long, and in most places six Miles broad, and a hundred Fathoms deep about the middle of it, it makes a great Bay to the Westward, and there lies here two Islands called the *Borromean* Islands, that are certainly the loveliest spots of ground in the World, there is nothing in all Italy that can be compared to them, they have the full view of the Lake, and the ground rises so sweetly in them that nothing can be imagined like the Terraces here, they belong to two Counts of the *Borromean* family. I was only in one of them, which belongs to the head of the Family, who is Nephew to the famous Cardinal known by the name of St *Carlo* . . . The whole Island is a garden . . . and because the figure of the Island was not made regular by Nature, they have built great Vaults and Portica's along the Rock, which are all made Grotesque, and so they have brought it into a regular form by laying earth over those Vaults. There is first a Garden to the East that rises up from the Lake by five rows of Terrasses, on the three sides of the Garden that are watered by the Lake, the Stairs are noble, the Walls are all covered with Oranges and Citrons, and a more beautiful spot of a Garden cannot be seen: There are two build-ings in the two corners of this Garden, the one is only a Mill for fetching up the Water, and the other is a noble Summer-House all Wainscotted, if I may speak so, with Alabaster and Marble of a fine colour inclining to red, from this Garden one goes in a level to all the rest of the Alleys and Parterres, Herb-Gardens and Flower-Gardens, in all which there are Varieties of Fountains and Arbors, but the great Parterre is a surprizing thing, for as it is well furnished with Statues and Fountains, and is of a vast extent, and justly scituated to the Palace, so at the further-end of it there is a great Mount, that face of it that looks to the Parterre is made like a Theatre all full of Fountains and Statues, the height rising up in five several rows . . . and round this Mount, answering to the five rows into which the Theatre is divided, there goes as Many Terrasses of noble Walks, the Walls are all as close covered with Oranges and Citrons as any of our Walls in *England* are with Laurel: the top of the Mount is seventy foot long and forty broad, and here is a vast Cestern into which the Mill plays up the water that must furnish all the Fountains . . . The freshness of the Air, it being both in a Lake and near the Mountains, the fragrant smell, the beautiful Prospect and the delighting Variety that is here makes it such a habitation for Summer that perhaps the whole World hath nothing like it.

BISHOP GILBERT BURNET, 1686

No Ideas you could form in the Winter can make you imagine what *Twickenham* . . . is in this warmer Season. Our River glitters beneath an unclouded Sun, at the same time that its Banks retain the Verdure of Showers. Our Gardens are offering their first Nosegays; our Trees, like new Acquaintances brought happily together, are stretching their Arms to meet each other, and growing nearer and nearer every Hour: The Birds are paying their thanksgiving Songs for the new Habitations I have made 'em: My Building rises high enough to attract the eye and curiosity of the Passenger from the River, where, upon beholding a Mixture of Beauty and Ruin, he enquires what House is falling, or what Church is rising?

ALEXANDER POPE, 1720

From the River Thames you see thro' my arch up a Walk of the Wilderness to a kind of open Temple, wholly compos'd of Shells in the Rustic Manner; and from that distance under the Temple you look down thro' a sloping Arcade of Trees, and see the Sails on the River passing suddenly and vanishing, as thro' a Perspective Glass. When you shut the Doors of the Grotto, it becomes on the instant, from a luminous Room, a *Camera obscura*; on the Walls of which the Objects of the River, Hills, Woods, and Boats, are forming a moving Picture in their visible Radiations: And when you have a mind to light it up, it affords you a very different Scene; it is finish'd with Shells interspersed with Pieces of Looking-glass in angular forms; and in the Cieling [*sic*] is a Star of the same Material, at which when a Lamp (of an orbicular Figure of thin Alabaster) is hung in the Middle, a thousand pointed Rays glitter and are reflected over the Place.

ALEXANDER POPE, 1725

An anonymous visitor who clearly experienced the authentic Gothic frisson recorded his impression of Pope's grotto:

Cast your Eyes upward, and you half shudder to see Cataracts of Water precipitating over your head, from impending Stones and Rocks, while Salient Spouts rise in rapid Streams at your Feet: Around, you are equally surprized with flowing Rivulets and rolling Waters, that rush over airey Precipices, and break amongst Heaps of ideal Flints and Spar. Thus, by a fine Taste and happy Management of Nature, you are presented with an undistinguishable Mixture of Realities and Imagery.

Newcastle General Magazine, 1748

When I saw you [William Shenstone], you talked of giving a short History of false taste: I can furnish you with one or two real facts that are not unpleasant. Last year died a Mr Weaver, who had a Seat at Morville near Bridgenorth and who was possessed by the very demon of Caprice: He came into possession of an Old Mansion that commanded a fine view down a most pleasing Vale, he contrived to intercept it by two straight rows of Elms that ran in an oblique direction across it, and which led the Eye to a pyramidal Obelisk composed of one single board set up endways and painted by the Joiner of the Village: this obelisk however was soon removed by the first puff of wind.

In view of one of his windows grew a noble large, Spreading Ash, which tho' the spontaneous gift of Nature, was really a fine object: and by its stately figure and chearful Verdure afforded a most pleasing relief to the Eye; you will stare when I tell you that Mr W. had this Tree painted *white*, – leaves and all: it is true the leaves soon fell off, and the tree died, but the Skeleton still remains, as a Monument of its owner's Wisdom and Ingenuity.

BISHOP THOMAS PERCY, 1760

I feasted upon grapes and ortolans with great edification; then walked to one of the bridges across the Arno, and surveyed the hills at a distance, purpled by the declining sun. Its mild gleams tempted me to the garden of Boboli, which lies behind the Palazzo Pitti, stretched out on the side of a mountain. I ascended terrace after terrace, robed by a thick underwood of bay and myrtle, above which rise several nodding towers, and a long sweep of venerable wall, almost entirely concealed by ivy. You would have been enraptured with the broad masses of shade and dusky alleys that opened as I advanced, with white statues of fauns and sylvans glimmering amongst them; some of which pour water into sarcophagi of the purest marble, covered with antique relievos. The capitals of columns and ancient friezes are scattered about as seats. On these I reposed myself, and looked up to the cypress groves which spring above the thickets; then, plunging into their retirements, I followed a winding path, which led me by a series of steep ascents to a green platform overlooking the whole extent of wood, with Florence deep beneath, and the tops of the hills which encircle it jagged with pines; here and there a convent, or villa, whitening in the sun. This scene extends as far as the eye can reach. Still ascending I attained the brow of the eminence, and had nothing but the fortress of Belvedere, and two or three open porticoes above me. On this elevated situation, I found several walks of trellis-work, clothed with luxuriant vines, that produce, to my certain knowledge, the most delicious clusters. A colossal statue of Ceres, her hands extended in the act of scattering fertility over the country, crowns the summit, where I lingered, to mark the landscape fade, and the bright skirts of the western clouds die gradually away. Then descending alley after alley, and bank after bank, I came to the orangery in front of the palace, disposed in a grand amphitheatre, with marble niches relieved by dark foliage, out of which spring tall aerial cypresses. This spot brought the scenery of an antique Roman garden full into my mind. I expected every instant to be called to the table of Lucullus hard by, in one of the porticoes, and to stretch myself on his purple triclinia; but waiting in vain for a summons till the approach of night, I returned delighted with a ramble that had led my imagination so far into antiquity.

WILLIAM BECKFORD, 1782–3

Because I take the garden I have named [Moor Park, Hertford-shire] to have been in all kinds the most beautiful and perfect, at least in the figure and disposition that I have ever seen, I will describe it for a model to those that meet with such a situation, and are above the regards of common expence. It lies on the side of a hill, upon which the house stands, but not very steep. The length of the house, where the best rooms and of most use or pleasure are, lies upon the breadth of the garden; the great parlour opens into the middle of a terras gravel-walk that lies even with it, and which may lie, as I remember, about three hundred paces long, and broad in proportion; the border set with standard laurels and at large distances, which have the beauty of orange-trees out of flower and fruit. From this walk are three descents by many stone steps, in the middle and at each end, into a very large parterre. This is divided into quarters by gravel-walks, and adorned with two fountains and eight statues in the several quarters. At the end of the terras-walk are two summer-houses, and the sides of the parterre are ranged with two large cloisters open to the garden, upon arches of stone, and ending with two other summer-houses even with the cloisters, which are paved with stone, and designed for walks of shade, there being none other in the whole parterre. Over these two cloisters are two terrasses covered with lead and fenced with balusters; and the passage into these airy walks is out of the two summer-houses at the end of the first terras-walk. The cloister facing the south is covered with vines, and would have been proper for an orange-house, and the other for myrtles or other more common greens, and had, I doubt not, been cast for that purpose, if this piece of gardening had been then in as much vogue as it is now.

From the middle of this parterre is a descent by many steps flying on each side of a grotto that lies between them, covered with lead and flat, into the lower garden which is all fruit-trees ranged about the several quarters of a wilderness which is very shady; the walks here are all green, the grotto embellished with figures of shell-rock-work, fountains, and water-works. If the hill had not ended with the lower garden, and the wall were not bounded by a common way that goes through the park, they might have added a third quarter of all greens; but this want is supplied by a garden on the other side the house, which is all of that sort, very wild, shady, and adorned with rough rock-work and fountains.

HORACE WALPOLE, *c.* 1750–60

I have just finished a new walk far superior in grandeur & variety to any I had. It even surpasses *my* expectations. You will be delighted when you see it for it is indeed beautiful & completes the circuit on each side [of] the river. I mean to have a swing bridge of Chains from Rock to Rock, so do not be surprised if you read in some Tourist of my having caused the death of several by fear or drowning.

THOMAS JOHNES, 1804

Of prospect I have a rich profusion and offering itself at every point of the compass. Mountains distant & near, smooth & shaggy, single and in ridges, a little river hiding itself among the hills so as to shew in lagoons only, cultivated grounds under the eye and two small villages. To prevent a satiety of this is the principal difficulty. It may be successively offered, & in different portions through vistas, or which will be better, between thickets so disposed as to serve as vistas, with the advantage of shifting the scenes as you advance on your way.

THOMAS JEFFERSON, 1806 (pub. 1987)

In the town [Charleston, Virginia], behind their high walls, grew oleanders and pomegranates, figs and grapes, and bulbs brought from Holland, jonquils and hyacinths. The air was fragrant with the sweet olive, myrtle and gardenia. There were old-fashioned roses; the cinnamon, the York and Lancaster, the little white musk, and the sweet or Damascus. The glowy-leaved Cherokee clothed the walls with its great white disks, and was crowded by jasmine and honeysuckle. The lots were so large, often a square or a half square, that the yard, stables and servants quarters were quite separate from these pleasant places, where, according to the fashion of the time, there were arbours, in which the gentlemen smoked their pipes, and the ladies took their 'dish of tea' of an afternoon.

HARRIOTT HORRY RAVENEL, 1906

What a soft, fresh delicious evening it was! He had quitted his carriage at the lodge, and followed it across the small but picturesque park alone and on foot. He had not seen the place since childhood – he had quite forgotten its aspect. He now wondered how he could have lived anywhere else. The trees did not stand in stately avenues, nor did the antlers of the deer wave above the sombre fern; it was not the domain of a grand seigneur, but of an old, long-established English squire. Antiquity spoke in the moss-grown palings, in the shadowy groves, in the sharp gable-ends and heavy mullions of the house, as it now came in view, at the base of the hill covered with wood – and partially veiled by the shrubs of the neglected pleasure ground, separated from the park by the invisible ha-ha. There gleamed in the twilight the watery face of the oblong fish-pool, with its old-fashioned willows set at each corner – there, grey and quaint, was the monastic dial – and there was the long terrace walk, with discoloured and broken vases, now filled with orange or the aloe, which, in honour of his master's arrival, the gardener had extracted from the dilapidated green-house. The very evidence of neglect around, the very weeds and grass on the half-obliterated road, touched Maltravers with a sort of pitying and remorseful affection for his calm and sequestered residence.

BULWER LYTTON, 1837

We live on the ground-floor, and our windows open on to a narrow terrace with a low stone parapet, from which one can throw a stone down into the river. The Karkaría makes a sharp bend just above Shustar, round what looks like the most beautiful park, a level greensward with immense dark green shady trees, standing as if planted for ornament. Here we sit, and late in the evening and early in the morning I see a pair of pelicans swimming or flying below. The terrace communicates with the garden, which is gay with poppies, pink and lilac and white, in full bloom. There is a little tank, and a row of stunted palm-trees, where rollers, green and blue birds like jays, sit, while swifts dart about catching musquitoes and flies, only a few hundred, alas, out of the millions that torment us. For there is no rose without a thorn, nor is this lovely kiosk and garden full of blooming poppies without its plague. The flies and musquitoes are maddening, and to-day the heat of summer has burst upon us.

LADY ANNE BLUNT, 1881

The garden was one of those old-fashioned paradises which hardly exist any longer except as memories of our childhood: no finical separation between flower and kitchen-garden there; no monotony of enjoyment for one sense to the exclusion of another; but a charming paradisiacal mingling of all that was pleasant to the eye and good for food. The rich flower-border running along every walk, with its endless succession of spring flowers, anemones, auriculas, wall-flowers, sweet-williams, campanulas, snap-dragons, and tiger-lilies, had its taller beauties such as moss and Provence roses, varied with espalier apple-trees; the crimson of a carnation was carried out in the lurking crimson of the neighbouring strawberry beds; you gathered a moss-rose one moment and a bunch of currants the next; you were in a delicious fluctuation between the scent of jasmine and the juice of gooseberries. Then what a high wall at one end, flanked by a summer-house so lofty, that after ascending its long flight of steps you could see perfectly well that there was no view worth looking at; what alcoves and garden-seats in all directions; and along one side, what a hedge, tall, and firm, and unbroken like a green wall!

GEORGE ELIOT, 1858

It was a quaint old place, enclosed by a thorn hedge so shapely and dense from incessant clipping that the mill boy could walk along the top without sinking in – a feat which he often performed as a means of filling out his day's work. The soil within was of that intense fat blackness which is only seen after a century of constant cultivation. The paths were grassed over, so that people came and went without being heard. The grass harboured slugs, and on this account the miller was going to replace it by gravel as soon as he had time; but as he had said this for thirty years without doing it, the grass and the slugs seemed likely to remain.

THOMAS HARDY, 1880

The outskirts of the garden in which Tess found herself had been left uncultivated for some years, and was now damp and rank with juicy grass which sent up mists of pollen at a touch; and with tall blooming weeds emitting offensive smells – weeds whose red and yellow and purple hues formed a polychrome as dazzling as that of cultivated flowers. She went stealthily as a cat through this profusion of growth, gathering cuckoo-spittle on her skirts, cracking snails that were underfoot, staining her hands with thistle-milk and slug-slime, and rubbing off upon her naked arms sticky blights which, though snow-white on the apple-tree trunks, made madder stains on her skin . . .

THOMAS HARDY, 1891

He [Mr Boythorn] lived in a pretty house, formerly the parsonage house, with a lawn in front, a bright flower-garden at the side, and a well-stocked orchard and kitchen-garden in the rear, enclosed with a venerable wall that had of itself a ripened ruddy look. But, indeed, everything about the place wore an aspect of maturity and abundance. The old lime-tree walk was like green cloisters, the very shadows of the cherry-trees and apple-trees were heavy with fruit, the gooseberry-bushes were so laden that their branches arched and rested on the earth, the strawberries and raspberries grew in like profusion, and the peaches basked by the hundred on the wall. Tumbled about among the spread nets and the glass frames sparkling and winking in the sun there were such heaps of drooping pods, and marrows, and cucumbers, that every foot of ground appeared a vegetable treasury, while the smell of sweet herbs and all kinds of wholesome growth (to say nothing of the neigbouring meadows where the hay was carrying) made the whole air a great nosegay. Such stillness and composure reigned within the orderly precincts of the old red wall that even the feathers hung in garlands to scare the birds hardly stirred; and the wall had such a ripening influence that where, here and there high up, a disused nail and scrap of list still clung to it, it was easy to fancy that they had mellowed with the changing seasons and that they had rusted and decayed according to the common fate.

CHARLES DICKENS, 1852–3

The design of the rockwork [of Hoole House, Chester] was taken from a small model representing the mountains of Savoy, with the valley of Chamouni: it has been the work of many years to complete it, the difficulty being to make it stand against the weather. Rain washed away the soil, and frost swelled the stones: several times the main wall failed from the weight put upon it. The walls and the foundation are built of the red sandstone of the country; and the other materials have been collected from various quarters, chiefly from Wales; but it is now so generally covered with creeping and alpine plants, that it all mingles together in one mass. The outline, however, is carefully preserved; and the part of the model that represents 'la Mer de Glace' is worked with grey limestone, quartz, and spar. It has no cells for plants: the spaces are filled up with broken fragments of white marble, to look like snow; and the spar is intended for the glacier. We may add that it is equally impossible to create anything like it by mere mechanical means. There must be the eye of the artist presiding over every step; and that artist must not only have formed an idea of the previous effect of the whole in his own mind, but must be capable of judging of every part of the work as it advances, with reference to that whole. In the case of this rockwork, Lady Broughton was her own artist; and the work which she has produced evinces the most exquisite taste for this description of scenery. It is true it must have occupied great part of her time for six or eight years past; but the occupation must have been interesting, and the result, as it now stands, must give Her Ladyship the highest satisfaction.

The rockwork is planted with a selection of the most rare and beautiful alpines, particularly with all the close-growing kinds; each placed in a nidus of suitable soil, and the surface protected from the weather by broken fragments of stone, clean-washed river gravel, the debris of decaying rock, moss, or other suitable substances; according as the object was to retain moisture; to evaporate moisture, in order to prevent the plants from damping off; to increase the heat, in which case dark fragments of stone are used; or to diminish it, which is effected by the employment of white pebbles, which, by reflecting the light and heat, keep the ground cool.

JOHN CLAUDIUS LOUDON, 1831

*S*trathfieldsaye, *his Grace the Duke of Wellington.* – We entered this noble park by an avenue a mile in length of elms, of a broader-leaved kind than the common English elm, and forming a tree of less altitude. The surface over which this avenue passes is undulating, which detracts somewhat from its first impression; but, as it is found to increase in length as we advance along, the sentiment of grandeur is recalled, and by prolongation is even heightened. We expected the surface of the grounds to be flat, but were agreeably surprised to find a gentle hollow running through them in the direction of the length of the park, in the bottom of which hollow is the river Loddon, widened, and otherwise heightened in effect. The park is as well wooded as could be desired, with trees of all ages and sizes, but chiefly with old oaks and elms. The avenue of elms terminates at a short distance from the house, where the pleasure-ground commences on the left, and a plantation continues to the kitchen-garden and stable offices to the right.

We met Mr Cooper, the very polite and well-informed gardener, at the commencement of the pleasure-grounds, and walked round them and the kitchen-garden with him, leaving the place afterwards by the London approach, which branches from the avenue in a winding direction at about two thirds of its length from the house. The pleasure-ground is of very limited extent, and perfectly flat; but it contains some very fine specimens of cedars, larches, Weymouth pines, spruce firs and other foreign trees.

Mr Cooper forces 25 sorts of figs; the duke, like ourselves, esteeming that fruit beyond all others. Some trees which Mr Cooper has removed from a wall to a forcing-house are 45 years old. There is a vinery stocked with plants 6 years old, producing an excellent crop. Mr Cooper has invented a very excellent utensil for sending cut flowers to London, or to any distance, without injury: It is simply a cylinder of tin, or of any other suitable material, of 3 or 4 feet in length, and 8 or 9 inches in diameter. In the centre of this is a cylinder of tin of an inch in diameter, which fits into sockets in the bottom and in the lid. Round this small cylinder the flowers are tied as they are upon a maypole; the pole so charged is inserted in the socket in the bottom, then the tube is filled with water, and corked, and the lid put on, in which is a socket, which embraces the tube. The case may now be sent to any distance, the water keeping the flowers cool and fresh. Mr Cooper informed us that the Duke of Wellington gave him some chestnuts which he had received from America, gathered from the tree which General Washington planted with his own hands, and from which (more fortunate than we have been, though we have received chestnuts three times from the same tree, once from Mrs Seaton of Washington, and twice from Dr Mease of Philadelphia,) he has raised three or four plants. We should be curious to know on what principle these chestnuts were sent to the Duke of Wellington; not that the merits of the latter general are at all less than those of the former, because we believe that the actions of all men are the joint results of their organization and the circumstances in which they are placed; but that we should like to know the feelings of the sender, and whether he was a Briton or an American. We have always had a great respect for the straight-forward character of the Duke of Wellington, and a profound admiration of General Washington; but with reference to all that is essentially grand in human nature, we have never for a moment placed the former on a par with the latter. As to the Duke of Wellington's private character as a husband and a master, all that we have heard at Strathfieldsaye and its neighbourhood places him, and also the late duchess, very high in our estimation. A spot was pointed out to us where it was intended to erect the new palace, the model for which, we were informed, is in one of the rooms of the present house. We hope it is not a frigid compilation in the Grecian or Roman manner. We should wish to see a magnificent pile in the old English or Italian style.

The charger which the duke rode at Waterloo is kept in a paddock adjoining a small flower-garden, from which the late duchess used frequently to feed him with bread from her own hands. During the battle, the duke was on this horse 15 hours, without once dismounting, and it has never been ridden since that day. It is a small chestnut horse, slightly made, and, as it was quite a colt at the time of the battle, it is wonderful how its strength was equal to the excessive fatigue it must have undergone. There is a proverb in some parts of England, that a chestnut horse is always a good one, and that it will always do more work than any horse of the same size, of any other colour, and this horse seems to furnish an illustration of its truth.

JOHN CLAUDIUS LOUDON, 1833

In the case of an old garden, mellowed by time, we have, I say, to note something that goes beyond mere surface-beauty. Here we may expect to find a certain superadded quality of pensive interest, which, so far as it can be reduced to words, tells of the blent influences of past and present, of things seen and unseen, of the joint effects of Nature and Man. The old ground embodies bygone conceptions of ideal beauty; it has absorbed human thought and memories; it registers the bequests of old time. Dead men's traits are exemplified here. The dead hand still holds sway, the pictures it conjured still endure, its cunning is not forgotten, its strokes still make the garden's magic, in shapes and hues that are unchanged save for the slow moulding of the centuries. *Really*, not less than metaphorically, the garden-growths do keep green the memories of the men and women who placed them there, as the flower that is dead still holds its perfume.

JOHN D. SEDDING, 1890

'I could *live* in it,' he said.

It was a little plot of ground, some fifteen feet square, abutting on the high-road, one of a succession of cottage-gardens, all of them of pretty much the same size, but each having a representative character of its own, and better or worse cultivated, more or less affectionately tended, according to the disposition, taste, and energy of the owner. This one was very formal, – but, indeed, from the narrowness of their territory, they necessarily all had that characteristic, – but noticeably neat and lovingly ordered. Its main ornament was a giant *Echeveria*, which drew my attention, certainly not by reason of its loveliness, but rather by the heartiness of its growth, somewhat surprising in a comparatively tender species exposed to all the chances of the year. Round it, at carefully-calculated distances, were geraniums, calceolarias, ageratums, some ten-week stocks, – everything, in fact, that you have a right to look for in a highly-respectable enclosure. The man I had addressed was a mechanic, employed in some neighbouring railway works, and he evidently treated his spruce little plot like a machine, which ought never to be out of gear. He had cast aside the dress of his daily occupation, smartened himself up, and put on his best attire, as he always did when about to work among his flowers – as though the tidiness he exacted from them reacted on himself, and compelled him, in turn, to be spick-and-span when in their superior company. I had stopped to compliment him on the assiduity with which he cultivated his bit of ground, and for friendliness' sake observed that he must indeed be fond of it. Then came the emphatic answer –

'I could *live* in it.'

I suppose I smiled; for a whole life passed on a piece of earth fifteen feet square, part of which is dedicated to a gravel path, seems a somewhat narrow existence. But, after all, what is narrow?

ALFRED AUSTIN, 1894

Frenchmen at the end of the eighteenth century became inclined towards rustic simplicity. They ornamented their pleasure grounds with hermitages à la Rousseau; picnicked on milk and fresh fruit from bowls of gold (or if the hostess was Marie Antoinette from dishes of pink Sèvres porcelain the exact shape and size of her own perfect breasts), the gentlemen dressing themselves as shepherds, the ladies as shepherdesses, or sometimes the other way round to add a transvestite piquancy to their attic ruralizing.

The new style garden, where all these pastoral high jinks went on, was called a *bagatelle* or *folie* – the perfect illustration being, of course, Bagatelle.

Ah, Bagatelle! It is one of the very few gardens in the world that can move me to make a spontaneous interjection mark in print, for I do not care to risk making a fool of myself by misusing stark work-aday words and expressions like *Woe! Alas! Ah, me!* and *God forbid!* which less inhibited writers appear to handle with such dexterity. But Bagatelle is different. It releases my pen. I can write *Ah* as deftly as any Edwardian melodramatist; and if there was in our language a single word which suitably summed up the gasp, then little puff, and long, long sigh which Bagatelle draws from an admirer, then I should write that too. And no wonder. It is a marvellous part of the Bois de Boulogne where for a wager the Comte d'Artois had a Pavilion built and pleasure grounds laid out all in a matter of seven weeks – the workmen being stirred to heroic feats by a never-ceasing supply of wine and the music of bagpipes and barrel organs.

Ah, Bagatelle! – where an English garden, a Chinese garden, and a French garden were designed to perfection, by a Scotsman; which survived the excesses of the Terror and became a promenade and park where citizens were entertained by stately courtesans, balloonists and brass bands, though they were instructed *à ne toucher à rien, à s'y comporter avec décence, à peine d'être arrêté.*

Ah, Bagatelle! To change your face and shape like a maturing woman and lose, in the Second Empire, your hermitages, picnic grottoes and rustic *tempietti* but gain, in compensation, that exquisite Kiosque de l'Impératrice, and become a repository of such remarkable garden sculptures that duelling was discouraged in case half-spent bullets chipped the statuary.

Bagatelle! Bagatelle! Where Quentin Bauchart's Rose Garden remains so charming that it works an annual miracle at the International Competition of New Roses by making Rose traders appear to be less brash and money-grubbing than they are at Chelsea.

TYLER WHITTLE, 1969

In Italy, too, flowers grow and multiply in a way unknown to England, and it is wise to take advantage of Nature's bounty.

As the cabbage and artichokes of our prepossessors gave way to the daffodils, anemones, and tulips mostly brought from Holland, the produce of these bulbs yearly made demand for greater space. Small islets of foxgloves or columbines or larkspurs spread themselves into continents, and a splash of Love in the Mist flowed over into a sea of blue. The vigour, too, of the plants that love the soil is so great that to reduce them and their groups to the dimensions that are observed in what is called a well-kept garden would be to restrain their nature, and we mostly prefer to let them ramp. As much as possible we give Nature her head, and when she is ridden it is with the lightest snaffle.

The result is that from early spring to late autumn we see a mass of bloom. The daphne tells us spring has come, and we are snow white with Marinelli, the only cherry which will stand our salt air and soil. Then pink with peach blossom, sweet with lilac, gay with may, Forsythia, Deutsia, Spirea, Weigelea, and Azalea. A laburnum, the only one we could persuade to grow, reminds one of the Derby, and wistaria disputing with roses the clothing of our cottages, climbs to the top of tall trees to deck with flowers the leaves of Chionanthus. A little later Pittosporum with shining dark foliage and white intensely daphne-scented flowers, quits shrub size to figure as a tree – one I know reaches the second storey of a neighbouring palace – and Rhincospernum hides the stone garlands of our modest Venus, runs over the Faun that stands near by, and helps to hide a cottage front. Whilst the bower that stretches between the statues is a purple glory of Clematis Jackmanni.

Before this the garden is white and gold with daffodils that, blossoming with the Marinelli, blow on to greet the roses – the Emperor, Empress, and Sir Watkin only yielding in size and beauty to the well-named Incomparable Sulphur Phœnix, with a flower larger and sweeter than the gardenia it resembles; and our dozen rose gardens are carpeted with tulips and anemones.

We have found it answer to mix standard roses with the bush ones, the shade of the tall plants giving some shelter to the dwarfer ones from the hot summer sun. For the same reason it has succeeded extremely well to plant tulips and anemones and even strawberries among the roses. Of course the ground plants are planted thinly. And a rose-bed carpeted with Cardinal and Rose de Nice, or with the Caen or Chrysanthemum Anemones, is very lovely. It is delightful, too, to pick one's strawberries and cut one's tea rose from the same bed.

When the tulips are nearly over the Iris come in. Of these we have a great profusion. Purple, pale blue, and white, bronze and yellow, Florentine, English and German, Spanish and Japanese. In a long border that joins one side of the cherry orchard large groups of different kinds are growing, and when in flower they so fill the eye that as one looks up the border more than a hundred yards long, leading to the square of the cherry orchard, it seems one continuous mass of their bloom. Round three sides of this orchard there is little else except some lilies of the valley, the offshoots as it were of the fourth side, which is filled with them growing beside and under cabbage roses.

These delicious so-called lilies make full return for our love and small demand upon our care. It is an amusement for young women in the spring to pick them in large bunches, or rather to pull them, for the lily flower stalk should be pulled out of its socket, never broken. To favour this pastime and for ourselves, we have four or five plots some ten to fifteen yards long in the transverse borders; the sites chosen so as to prolong as much as possible the flowering season. It is necessary to change a bed in rotation nearly every year, for they grow so thick that the leaves suffocate the flowers. The first year they do little, the next two or three they bloom profusely, and the next more poorly. The double flowering plants, less social, require more space. They seem to do best dispersed about in small colonies, or singly, in places generally chosen by themselves, or moved to, aided by accident or perhaps by birds.

In May, early and mid, comes our great show, the roses. We have very many kinds, but we love best those that love us best, and are rather fond of putting a good many of any of the varieties that blow most freely in a mass, trying as to place to suit their idiosyncrasies. A large group of Comte de Paris, or Papa Gontier, in the shade they seek; or of Beauté Inconstante, Madame Jules Grolez, Maria Immaculata, in the sun they revel in. Further, there are so many roses, such as Gustave Regis, Souvenir de Catherine Guillot, Madame Eugène Resal, and her elder sister, Madame Laurette Messimy, Comtesse Riza du Parc, Caroline Testout, Madame Falcot, and the old Malmaison, and that sweetest of roses, La France, which, seeming to the climate born, give us flowers in such rich masses that we rather give up trying to grow the more fickle or contradictory, beautiful as they may be. Madame Hoste is unkind, but I hope to overcome her coyness. La Marque, Devoniensis, and Niphetos will have none of us. I am sorry, but as from April generally to Christmas, and sometimes to the Russian New Year's Day, we can cut from thousands of plants of hundreds of varieties that thrive with us, we may well be and are content.

FREDERICK EDEN, 1903

In all the world there is no place so full of poetry as that Villa d'Este which formalist and naturalist united to decry. Driving past the little Temple of Vesta, high above the seething cauldron of the Anio, one is admitted through vaulted corridors and deserted chambers where faded frescoes moulder on the wall to a stairway overhanging the garden. And the garden that lies in the abyss below, terrace after terrace looking out upon wooded mountain flank and far mysterious plain – surely Time has forgotten these giant cypresses which lift from the gulf dark pinnacles of night, great rugged, gloomy-verdured spires; surely it is the garden of a dream? Behind one like a cliff rises a palace of romance, vast, august, austere; a palace over which in a far-off age some mighty magician has thrown an enchanting spell of sleep. Sleep and forgetfulness brood over the garden, and everywhere from sombre alley and moss-grown stair there rises a faint sweet fragrance of decay.

SIR GEORGE SITWELL, 1909

I am not fond of gardens or borders devoted to one colour; but if ever I were tempted to make one it would be yellow in all the frank and pleasant tones from cream and buff and the bright butter yellows through apricot to the tawny ochreous shades, reaching now and then to flame. Not all blue flowers may be safely used in each other's company and but few pinks unless they are of the same scale; but all yellow flowers, like the light of which they seem to be fashioned, blend and combine or flash back at each other with never a jar to the most sensitive eye. They are the sunshine of the garden, and it is a pleasant fact that yellow flowers are more plentiful than any others and that

from the time of the delicate radiances of spring to the flaring up of
autumn's beacons their illumination is undimmed.

LOUISE BEEBE WILDER, 1918

Already in the summer of 1915, small and modest gardens began
to appear beneath the windows of a number of the barracks [at
Ruhleben prisoner-of-war camp]; and the suspicious whispered that
this was a German trick – that the Germans wanted to photograph
these gardens in order to prove to the world that their Ruhleben
prisoners were being pampered. But that was a mistake. The idea of
the gardens had originated with the gardeners – men whose sole
desire was to introduce a little beauty into their surroundings. Some
of them grew shrubs in order to hide the barbed wire. The sailors of
Barrack VIII made a rose-garden, and were very proud of it. Confidence was gradually established that the military authorities would
not interfere.

Then, after the gardens had become to bloom, the Horticultural
Society was founded. Its first president was Mr Warner of the Dunlop
Tyre Company, and its subscription was one mark a year . . . They had
already procured the use of a strip of ground behind the wash-house,
on which they had erected a few frames, in which to prepare plants for
the coming season; and in January, 1917, they raised the question:
Might they not be allowed to turn the unused half of the race-course
into a kitchen garden?

Negotiations to that end were opened with the military authorities.
Their sanction obtained, a contract was arranged with the proprietors, for the use of the space at a rent of one hundred marks per
month. Then volunteers were called for, tools were procured, the
land was tilled and manured, and seed was purchased in England. A

loan of £400 from the Captains' Committee launched the enterprise on its way; but it prospered so exceedingly that, at the end of the first season, not only had the whole of the loan been repaid, but a substantial reserve was in hand for the operations of 1918. It was not the least useful of the camp undertakings; and there were times when it may well have seemed the most useful of them. It meant that fresh vegetables were on sale at nominal prices in the Ruhleben canteen, at times when Berlin housewives were jostling each other for the privilege of buying them at exorbitant prices in the green-grocers' shops.

Naturally, none but our own men were allowed to purchase them. Naturally, too, the Germans tried to evade our regulations, just as we had evaded so many of theirs. 'During my absence from the camp,' notes Mr Powell, 'Captain Amelunxen demanded fifteen red cabbages for the officers' mess, and also gave an order that all outer cabbage leaves must be delivered to serve as cattle fodder. This at a time when red cabbages were hardly procurable for love or money in Berlin. The Horticultural Society laid the matter before the captains; and the captains decided unanimously that, as the cabbages had been grown from seed supplied from England on the specific understanding that only the interned were to benefit, the officers' mess must go without its fifteen red cabbages. As for the leaves, he was told that he might have them in exchange for manure to be supplied by German cattle. On that condition they were duly handed over; but I have reason to believe that they were intercepted on their way to the cattle, and eaten.'

Cabbages, however, constituted only a small portion of the Ruh-leben garden produce. There was a potting shed, and a glass house, with steam-heating supplied from a boiler – the whole of the house and heating installation put up by our own men; and in it was pro-duced a really wonderful crop of melons and tomatoes. Flowers, too, were cultivated. Cut flowers, and flowers in pots, decorated most of the boxes; and we had our horticultural shows. Both Baron and Bar-oness Gevers and Ridder van Rappart of the Netherlands Legation always made a point of attending these exhibitions, and particularly admired a show of sweet peas, grown by Mr Roberts, who desired, as an expert, to demonstrate that sweet peas could be grown in Germany in spite of the dry climate.

JOSEPH POWELL AND FRANCIS GRIBBLE, 1919

In March of 1919 I had a wonderful opportunity to see the battle-fronts of Europe from Nancy to Ostend. A sadder, more appalling vision of destruction never was. Town after town was leveled to heaps of brick and dust; tree after tree was deliberately sawed off and left to rot. The grape-vines were pulled up, the fruit trees girdled, the land itself so shattered and upheaved that the gardener's first query was whether it could again bear crops before the lapse of many years.

We had left Amiens one Sunday morning, and passing Villers-Bretoneaux – where the Australian troops and some American engineers had made the stand that saved Amiens and the Western line – had gone through Hamelet, Hamel, Bayonvillers, Harbonnières, and Crepy Wood to Vauvillers. As the only woman in the party, I had been unanimously appointed in charge of the commissariat. It was noon when we reached Vauvillers. I chose a broken wall about fifty feet from the road as a good place on which to spread our luncheon. The car was stopped, the luncheon things were unpacked, and we picked our way over the mangled ground to the fragment of wall. As I passed around the end I came upon two peony plants pushing through the earth. Tears brimmed. I could not control them. Here had been a home and a cherished garden. As I stood gazing at the little red spears just breaking through the ground, a voice, apparently from the sky, inquired whether Madame would like a chair. Looking along the wall I saw the head of an old peasant woman thrust through a tiny opening. She smiled and withdrew, appearing a moment later with a chair. It was her only chair. She then brought forth her only cup and saucer, her only pitcher filled with milk, and offered us her only hospitality!

Joined now by her venerable husband, we listened to their story. The hiding of their few treasures, the burial of their bit of linen, their flight toward Paris, the description of the outrageous condition of the one room left for them to return to, made us burn with indignation. It was in her little garden that the peonies grew. The fruit trees and shrubs were gone, the neat garden walls were blasted into space, the many precious flowers were utterly destroyed. When she found that Madame, too, loved *les belles pivoines*, she urged me to take one of the only two roots she had left!

We went away leaving the old couple laden with supplies, and I gathered from every man in our party a heavy toll of tobacco for a farewell gift of comfort. I hope she has again a little garden, with all the peonies that it will hold.

ALICE HARDING, 1923

The Garsington pond proved to be, from the first, the magnetic point of the garden: tragic, comic and beautiful beyond Ottoline [Morrell]'s imaginings. In the first year the old cowman mysteriously drowned himself in it. Four years later, the painter Mark Gertler *nearly* drowned there. In between, a land girl called Lucy *pretended* to be drowning under the fascinated gaze of Asquith, the Prime Minister; and, finally, the big black boar fell into it.

A small weatherboard boat-house – a white-painted pavilion with classical pediment – was transported by the Morrells from Peppard (the Morrells' previous country house, near Henley) and re-erected at one end of the pond; from it, stone steps led down into the water. Bathing parties and boating parties were the order of the day. Summer visitors borrowed bathing costumes on arrival and ran down the grassy slope to the pond – a scene immortalized in one of the set-pieces of Lawrence's *Women in Love*. The water was terribly cold, and to Dora Carrington, the painter, unappetizing. 'I lie exhausted in the sun,' she wrote in the hot summer of 1916, 'after swimming in that cess-pool of slime.' Later, she took off her clothes and posed beside the pond for her photograph, along with Julian and Juliette and Ottoline herself – living statues among the stone.

But it was the beauty of the pond that mattered. The yew hedges duly went in at the end of 1915 (D. H. Lawrence helped Lady Ottoline plant some of them) and soon peacocks were trailing their tails along the grass walks beside the water. W. B. Yeats remembered them thus, strutting

> With delicate feet upon old terraces . . .
> Before the indifferent garden deities . . .
> (*Meditations in Time of Civil War*)

On the far side of the pond a second yew hedge was planted behind the first, making a dark *allée* down which both birds and people could process, past *claire-voies* or peep-holes cut in the outer hedge to give framed views of the wide landscape beyond. They walked there in moonlight, reciting Verlaine, or in the afternoons, locked in *tête-à-tête*. Mark Gertler tried persistently to paint it all: the pattern of right-angles, the verticals of statues, the shining light of the water barred by the deep reflections of trees. But when he set up his easel on a sunny weekend, perambulating guests continually bore down on him. It was like painting in the middle of a market place or pleasure ground, he said.

DEBORAH KELLAWAY, 1993

When I think of scented gardens I remember hers [Great-aunt Lancilla's] first and foremost, for although since those days I have seen many gardens, I do not think I have ever seen a pleasanter, homelier one. The house was Georgian, and the short drive to it was flanked on both sides by pollarded lime trees. I have only to shut my eyes to hear the hum of the bees now. The drive was never used by the household nor indeed by anyone who came on foot, for the shortest way from the village was through a gate leading from the road to a side door. The path was perfectly straight, and bordered on either side by very broad beds, and except in winter they were full of scent and colour. I can see the big bushes of pale pink China roses and smell their delicate perfume; I can see the tall old-fashioned delphiniums and the big red peonies and clumps of borage, the sweet-williams, the Madonna and tiger lilies and the well-clipped bushes of lads-love. Before the time of roses I remember chiefly the Canterbury bells and pyrethrums, and earlier still the edge nearest the path was thick with wallflowers and daffodils. I have never seen hollyhocks grow as they grew at the back of these borders, and they were all single ones, ranging from pale yellow to the deepest claret.

Beyond this path, on one side was the big lawn with four large and very old mulberry trees. As a child it frequently struck me that considering how small mulberries were compared to apples, plums and so forth, it was really little short of a miracle what a glorious mess one could get into with them in next to no time. Amongst the flowers Great-aunt Lancilla loved most were evening primroses. I have never since then seen a large border, as she had, given entirely to them. She used to pick the flowers to float in finger-bowls at dinner.

I can see the kitchen garden, too, with its long paths and espalier fruit trees and the sweet-peas grown in clumps, and they *were* sweet peas then, deliciously scented. And big clumps of gypsophila and mignonette, which everyone in those days grew to mix with the sweet-peas. There were great rows of clove carnations for picking, and never have I smelt any like them. Nor have I since tasted the like of the greengages which grew against the old wall. Is there anything quite so good as the smell and taste of a ripe greengage, picked hot in the sun? I can see the orderly rows of broad beans, lettuces, peas and scarlet runners and stout cabbages. The onions and 'sparrer grass' were the special pride of the old gardener's heart. I can see the well and hear the pleasant clinking sound of the bucket as it was let down.

I can see old Gregory attending to the bee-hives with the calm, gentle movements which characterize all experienced bee-keepers. He invariably talked to the bees when he was attending them, and one day when, as a small child, I was watching him I asked 'Do the bees understand what you are saying to them, Gregory?' 'Understand, Missie?' he replied. 'Just as much as horses and dogs and cattle; it stands to sense an' reason they do! An' sometimes I thinks they understan' more nor we do.'

And the raspberries and gooseberries! My Great-aunt had a favourite Aberdeen terrier, who, incredible though it may seem, loved ripe gooseberries. He used to sit up, as though he were begging, and eat them off the bush and wail aloud every few minutes whenever his nose was pricked.

I love to think of the huge beds of lily-of-the-valley, where one could gather and gather to one's heart's content for friends in the village. But my chief recollection of that kitchen garden is of roses. Cabbage roses and *La France* and *Gloire de Dijon* and 'Maiden's Blush', and if one gathered armfuls it seemed to make no difference. Those were the days when people filled their rooms with innumerable small vases of flowers, but my Great-aunt, who always went her own way entirely, loved to have big bowls of flowers everywhere, even in the passages of her house.

ELEANOUR SINCLAIR ROHDE, 1936

. . . the average cottager prefers some bloom in his garden all the year round. He likes to see the same flowers year after year in their wonted places, and but seldom succumbs to the suggestion in wireless talk or gardening magazine to try some new variety. Yellow wallflowers he plants always near the house, red farther off; thrift, catmint, pinks or saxifrage bordering his paths, stocks below the windows when the 'polyanthums' are over, tulips to rise stiffly from between rows of fading crocuses. But if he is unimaginative where arrangement is concerned, he has a child-like love of bright colour, and Nature so has it that whether he 'shows' at Bramelham Long Fair or not, at this time of the year his plot will most satisfy him. There are sweet peas, climbing on their rather too professional scaffolding; roses and clematis and honeysuckle and hollyhocks that vindicate even the most sentimental of pictures; bushes of lavender, overblown because they have already yielded enough and to spare for the linen cupboards, white alyssum to set off the many rich reds and golds – for gardens seem, like trees, to flame in a final brilliance before they wither into winter, and now they are afire with red-hot pokers, dahlias and oriental poppies, with crimson snapdragons and orange nasturtiums and marigolds.

HEATHER AND ROBIN TANNER, 1939

One other enduring pleasure, throughout the years, has been the making of our garden [La Foce]. In the year after our marriage, my American grandmother – somewhat startled to find herself, in mid-summer, in a house in which there was so little water, even in the baby's bathroom – presented us with the wonderful gift of a pipe-line which, leading from a spring in the beech-wood at the top of our hill (some six miles away) brought us our first abundant water-supply. It then became possible to plan, not only new bathrooms, but a kitchen-garden and flower-garden, which gradually grew, year by year, in proportion to our means and to the water available. First, at the back of the house, I made a small enclosed Italian garden: a stone fountain with two dolphins and a small lawn around it, and a few flower-beds edged with box. A couple of years later, we made another larger terrace, passing through two pillars of travertine with ornamental vases into a less formal flower-garden, with wide borders of flowering shrubs, herbaceous plants and annuals, big lemon-pots on stone bases, a shady bower of wisteria and banksia roses, and a paved terrace with a balustrade, looking down over the valley, on which we would dine on summer nights when, just before the harvest, the whole garden would be alight with fireflies and the air heavy with nicotiana and jasmine. On the walls, in the spring, grow great clumps of aubretia and alyssum and, later on, rhyncospermum and climbing roses, and the grass is edged with daffodils and irises. Some steps – for the whole garden is on a fairly steep hillside – lead up to an avenue of cypresses and a rose-garden, while a wide pergola winds round the hillside towards the woods. Finally, just before the war, we made another enclosed formal garden – designed, like the first, by our friend and architect Cecil Pinsent, with hedges of cypress and box and big trees of magnolia grandiflora, while the rest of the hill above has been gradually trans-formed into a half-wild garden with Japanese fruit-trees and Judas-trees, forsythia, philadelphus, pomegranates and single roses, long hedges of lavender and banks fragrant with thyme, mint and assynth, and great clumps of broom. Gradually, by experiment and failure, I learned what would or would not stand the cold winters and the hot, dry summer winds. I gave up any attempt, in my borders, at growing delphiniums, lupins or phlox, as well as many other herbaceous plants; and I learned, too, to put our lemon-trees, plumbago and jasmine under shelter before the winter. But roses flourish in the heavy clay soil, and so do peonies and lilies, while the dry hillside is where laven-der thrives – a blue sea in June, buzzing with the bees whose honey is flavoured with its pungent taste, which also, in the winter, not only scents our linen but kindles our fires. Every year, the garden grows more beautiful; even the war brought no greater destruction than the shelling of a few cypress trees. The woods were already carpeted, according to the season, with wild violets, crocuses, cyclamen, anemone *alpina*, and autumn colchicum, and among these I also managed to naturalize some other kinds of anemones, daffodils and a few scillas. But bluebells I have failed with, and the exquisite scarlet and gold tulips which grow in the fields round San Quirico, just across the valley, still stubbornly refuse to flower here . . .

A friend, who has stayed with us in recent years, once wrote to me: 'I sometimes think that your garden is like an allegory of life itself: one passes from the warm, sheltered house into the formal garden, with its fountain and flowers and intricate box hedges, then coasts the hillside under the pergola of vines. The view opens out on to tilled fields, the flowers become rarer; one passes into the path through the woods. Here it is darker; the wind stirs in the branches. A few steps more, walking uphill in the shade, and one has reached the still chapel, with those four stone walls around it.'

Up that path, when the time comes, we both hope to go.

IRIS ORIGO, 1970

In my experience of vicarage gardens, our own, so far as the trees, the shrubs, and the design went, was typical of hundreds. The house was rebuilt in the forties, and the garden took its shape between the forties and the sixties. Before it began its decline into wilderness, there was only one major alteration. Croquet became popular, all of a sudden, in the fifties, so about 1860 a croquet lawn was added below the drawing-room window. In the nineties my parents enlarged the croquet lawn for the more popular game of tennis. Except for a buddleia (buddleia was introduced into England so late as 1902)*, everything that grew in the garden was either an ancient favourite of English horticulture or some plant introduced in the eighteenth century or the first fifty or sixty years of the nineteenth. You entered the garden by a drive, through large double gates painted green (drives were a minor index of class and were essential to the new vicarage demesnes). Left and right there were laurels, plain and variegated. Then on one side a clean and healthy monkey puzzle, on the other two locust trees, or as we commonly call them, acacias, and two large clumps of pampas grass from the Argentine rising out of the turf, and a weeping ash. Under the acacias and the weeping ash grew daffodils of the wild type and snowdrops and crocuses. The drive, designed for the waggonette and the dog-cart, curved around; and a flower-bed began which contained much *Anemone hybrida* and much white and pink *Spiraea* and a clump of the shiny evergreen *Fatsia japonica*, popularly and wrongly called the castor-oil plant. This long bed or border was edged with box and was backed with variegated laurel and two finely grown Portugal laurels.

The drive widened before the front door, and became flanked with laurel, holly, and rhododendron.

Left and right of the door the plants again were typical. Out of the grass beneath one window (which opened to the study) grew clumps of the holly-leaved barberry and of the red-flowering currant, which the children of the house had named 'Ginger Muck' in tribute to the sour smell of the flowers. Against the wall nearby sprawled a jasmine. Half hiding the window of the dining-room, a tall 'syringa', a mock-orange, that is to say, had been placed very carefully, not too near, which would have made the scent of the flowers overpowering in the summer as the vicar and his family sat at the mahogany table in the dining-room, and not too far for the scent to drift in with the right intensity. A path led round from the drive, past the 'syringa', to a long terrace under the southern windows. It passed first of all between two of the most clerical of plants, the white-flowered, winter-flowering

* Actually earlier. (Ed.)

and evergreen laurustinus and a thicket of the Rose of Sharon. The more sombre and formal evergreens were now left behind. Roses of various heights and kinds were clumped against the wall, against the stucco, there were paeonies again, then more flower-beds edged with box (in which love-in-the-mist came up year by year), then at the end of the terrace flowering shrubs and trees, fuchsia and laburnum, weigelia, escallonia, lilac, and japonica. Gradually they grew so thick and high that one could no longer gaze over the terrace wall to the white triangular hills of the distant clay works. One pierced through these shrubs to a doorway in the wall through which the gravelled path went downward to the kitchen gardens and the orchard hidden decently out of sight, and so maintaining the emphasis on the paradise of delights, the gentleman's pleasure garden.

GEOFFREY GRIGSON, 1952

In the afternoon I moon about with Vita [Sackville-West] trying to convince her that planning is an element in gardening. I want to show her that the top of the moat-walk bank must be planted with forethought and design. She wishes just to jab in the things which she has left over. The tragedy of the romantic temperament is that it dislikes form so much that it ignores the effect of masses. She wants to put in stuff which 'will give a lovely red colour in the autumn'. I wish to put in stuff which will furnish shape to the perspective. In the end we part, not as friends.

HAROLD NICOLSON, 1946 (pub. 1966)

The Editor of *Gardening Illustrated* wants to do an article about our garden [Sissinghurst]. That is nice of him, and furthermore he wants to incorporate the article in a book he is doing for *Country Life*. It *is* funny, isn't it, that our own dear garden should be taking its place among the better known gardens of England? Oh, if only we could have the last twenty years all over again! We wouldn't make any change in the design, but I should like to go back and make a great many changes in the planting. Beastly garden.

VITA SACKVILLE-WEST, 1954

The friends with whom we were staying, and indeed had made the several-hundred-mile-journey to see, suggested we take a walk. The dark, black canopy of a wood beckoned. A meshed thicket of branches when more closely focused revealed a classic late-winter woodland scene. There was not a sign of life; even the evergreen ivies had taken on a dull, sullen, lacklustre appearance. The undergrowth of bare brambles was made even harder going by the debris from the gale. We clambered over many fallen trees; it was difficult to enjoy the surroundings because of the constant necessity to concentrate on the obstacles before us. The ground became damp, and then I noticed that a stream seemed to have formed, magically, before us. It led on, winding downhill, down towards the edge of the wood. Our friend announced, 'There's a stile over here. We can climb over from the wood and into the field beyond. It's privately owned land but I know the owners. I don't think they'll mind.'

It was inexplicable, but when I climbed over that stile out of the wood, I knew that something felt very different. I have a memory of tall hedges rising before me, but at that moment I stood beside what I, from pure affection, would call a pond; it had a too essentially rural quality to be described as a lake. The more formal-looking hedges mingled so well with their rural cousins that I was doubtful whether they did indeed belong to some more formal setting, as my inner instinct led me to believe.

I cannot now recall how I made my way through those formal hedges. A white-painted summer-house stood before a sheet of water which lay like a mirror on that still winter's day. Most vividly I recall something sheltering under the hedge, all encased in grey, which I deduced to be a statue, all wrapped up in its winter coat. Without knowing how, I then found myself transported to an orchard. It was like an orchard *ought* to be, I thought: it had an abundance of apple trees, but mixed with other fruit trees too, and there were signs of other life as well. Shrub roses, their green-, brown- and corky-coloured stems simply set in the grass as though they had chosen for themselves that very spot in which to grow. Some, like gentle fronds, I pictured with pale, delicate and fragile-petalled flowers; others, robustly guarding themselves with huge thorns, looked as though they bore larger, vivid, attention-seeking flowers. Climbing roses were entangled among some of the apple tree branches, and caught a slight breeze which ruffled such leaves as had over-wintered, while the old apple tree stood silent and unmoved.

The feeling within me that this orchard was . . . special . . . different . . . increased. It was like hearing a great piece of music for the first time, which stirs those deep inner feelings, awakens one to greater sensitivity, and finally grips one in a total flood of its own power. There was an aura of timelessness, despite the fact that this was a living creation.

My husband and our friends had vanished, and I was far away, totally lost in what I had found, utterly consumed by the atmosphere. Even in the dullness of a dead February day it bore all the signs of love and care: some Prometheus had created this and brought it to life, and it radiated an aura that called forth those who would recognize the call, to taste, and to see.

Emerging from my reverie, I remembered that our friend had said he knew the owners; I rushed to catch the others, to ask. Like Alice in a dream, I ran along a narrow pathway between two tall yew hedges, and the need for me to know became in that moment a need beyond all others. The yew parted – the garden's spirit had carried me to its very heart, and there was no need to ask my burning question: a mellowed red-brick Elizabethan tower rose before me, the bastion of English gardening, to whose steps I had been taken, seemingly out of nowhere. It was Vita's tower at Sissinghurst Castle.

JANE ALLSOPP, 1994

With this garden of ours, regarded as a whole, three things have contributed to our content. The first is the knowledge that we started it from the raw, the second that its development has progressed by certain clearly defined stages, and the last is a sense of satisfaction in that what we have achieved is our own creation. Faults it may have, indeed has, in plenty. Sins of commission and omission may from its depths upbraid us in contemplative hours. But blest beyond all with a natural setting that has all along been our 'fostering star', it is, to us at all events, good and fair as it now mellows from adolescence towards maturity.

Three stages have marked the garden's growth and in their inception and subsequent influence there seems to have been some divinity at work in the shaping of our ends. For the garden of the first period as well as of the second – each roughly covering in possession a decade or more – was planted-up and so far comfortably furnished in a garden sense before we acquired the additional portions, each of which were an extension of that already existing.

Thus to the original wooded slope through which we first blazed the trail, which meant doing little in particular beyond preserving the *status quo*, was added the glade, to be followed shortly by the dingle and its riverside walk to our annex at the south, so securing the main trough of our little valley. But there was still the northern end of the latter with its alluvial flat inviting an extension of the glade, its flanks of old oaks, its ancient water-mill and other appurtenances, the whole covering some two acres. Though from the first we had not been unaware that the addition of this remaining section was essential if ever we were to round off our little estate and make it a self-contained whole, exclusively embraced by the river on the one side and the road on the other, hopes of our ever being able to acquire it were too remote for serious consideration.

True, we oft-times felt as King Ahab felt when he regarded with covetous eyes the vineyard of Naboth. But our worthy neighbour was as obdurate over retaining his own as that sticky Jezreelite, and though we wanted that vineyard, wanted it badly sometimes, we put the desire from us and went our way, satisfied with our own. Even so, if destiny had tarried long, for nine years to be exact, it smiled upon us at the end, and one autumn day, when the world was trembling on the brink of war, the vineyard became our own. It was offered to us gladly and without reserve. It came along at an hour when we were more fully prepared to undertake it than we had hitherto been, more consciously aware of its necessity as a factor in carrying out our garden programme. It came along, too, at that moment in the year's annual round when nature was repeating in the russet and gold of oaks, in the lambent refulgence of beech, in the chequer of ivory that had invaded the bracken, the old old story of life's surrender that life might live, a story, it seemed, of our own garden progress. First the blade, then the ear and, at length, the full corn in the ear.

ARTHUR TYSILIO JOHNSON, 1950

That road sign, 'Heavy Plant Crossing' – yes, we know what it *means* – always suggests the possibility that some gigantic swede or vast turnip, out for a stroll, will rear up in our path. Too many science fiction films, of course, and an overheated imagination of rambling habit. But you can plant a kiss, or a thought, or a bulb. You can even transplant a heart. And there is the Hardy Plant Society, Hampshire branch, whose pleasant custom it is to adventure forth and look at gardens . . .

A lady I know well telephoned me early in April last year to ask me to come and provide as it were a little personal ground-cover the day the HPS were coming to visit her garden. I conjectured that it was my duty to get there early, to be discovered perhaps in rapt study of a tulip when the curtain went up. After all, who knew what the weather would do on 28 April? Hardy the society may be, but would they come in the teeth of a gale under a tenebrous sky with wildly flapping curtains of rain? – hardly that hardy, surely.

Well, it was a day of lupin-blue skies, no breeze at all and the temperature in the seventies. My bit of ground-cover was not needed, but I wondered what manner of person were the Hardy Planters. What made them, not so much tick, as sprout. I prowled, I eaves-dropped, I observed. Those who had brought their children had raised their little seedlings with care; they were old-fashioned children of the seen but not heard variety; and their dogs, too, comfortably asleep in their owners' cars, were models of canine deportment (mostly gold labradors, I noted); and their cars were of the sensible variety, and all the Volvos' back window sills had a copy of the Yellow Book wilting in the sunlight.

'Something horrid got at my hostas . . .'

'Oh, no . . .'

'But I sprayed them and hope they may yet come round.'

'Just look at those tulips!'

'I say!' Our attention swivelled to where a man with a zoom lens was taking mug shots of some white – but with green at the base of the cups – tulips. Not that the tulips were guilty of anything, except perhaps superbia. They soared like a Bruckner motet from out of a white marble urn. 'What are they called?' I asked a lady who looked as if she was being played by Joyce Grenfell. I really wanted to know.

'Viridiie', she said, giving each *i* full value. She smiled briefly – it was like a glimpse of an organ loft – white pipes – and turned to a friend who was being played by Athene Seyler. 'You must let me run you down for a look at Virginia's bog.'

'I knew,' conceded Athene, her face crumpling in recollection, 'she was planning a pond, but a bog-garden . . . I could let her have some king-cups, we could make a day of it.' They drifted, two nice tweed clouds, out of ear-shot, leaving me in a tumble-dryer of envy. Virginia's bog AND pond, king-cups, king-cups, king-cups. And would minnows dart through the Waterford crystal water? Would lucky Virginia have water hyacinths, I wondered. Humming 'There'll Always be an England', I decided to spy in another part of the garden.

No garden, I maintain, however small, should reveal itself to the single glance. Of course in a suburban garden, space is a factor, but the garden here was large, capacious enough to contain a great cliff of rhododendrons. Little gashes of colour, premonitory of the magenta and crimson flood to come, already showed on the dark green revetments. Then again, how many gardens could be home to what the Hardy Plant Society's account of the visit calls 'a truly magnificent weeping beech of great stature' without being swamped? That weeping beech has a sort of interior chamber which always puts me in mind of the Chapter House at Salisbury Cathedral. On the floor of the chamber retired hyacinths live happily, *and* tiny cyclamen and violets and primroses. The weeping beech is taller than many a parish church. It is said that Her Late Majesty Queen Mary once paused to inspect it; this may or may not be true, but if it is true then the royal curiosity is entirely sympathetic and understandable. As the account of the garden points out, Angela's garden is on part of the site of an old nursery. Which would probably account for a delightful old frowsty granny of a palm tree which always looks to me as if it's wearing a surgical stocking. Invalidish, yes, but viewed at the right season and from a flattering angle it makes you want to cry aloud 'We take the golden road to Samarkand', or 'Saddle my camel.'

But to return to my point about not being able to see into every corner at a glance. Espaliered apples and pears forming a tracery wall afford the possibility of assignations, plots and ambushes. You can explore too, if you wish, the caves at the foot of the cliffs of rhododendrons. Good gardens contain mysteries just as much as 'in a border out of the morning sun the shady greens had been cleverly lightened by a sprinkling of white honesty' – such features, clever ways with white honesty, are vital of course, but while enjoying them with one eye the other eye should be teased: 'what exactly is going on under the skirts of the weeping beech?'

Sipping coffee, crunching biscuits, the members drifted to and fro observing here a *Symphoricarpos orbiculatus* 'Foliis Variegatis', there a *Fremontodendron californicum* basking in the sun. Nice people, I felt; civilized. Gardening is such a sympathetic occupation, and brings out the best in people. Only a day or two later I was to visit Frogmore, and among the visitors to that august plot you could pick out the Hardy Planters, notebooks in hand, keen eyed, murmurous, from the people who, to be honest, might just as well have been in Disneyland. I seemed to feel the presence of Virginia of the Bog Garden, and *she* wouldn't be licking a cornet. Not that all people who garden are without stain. No, indeedy.

JOHN FRANCIS, 1991

As I entered my garden I removed the dark glasses I had been wearing, and halted in astonishment. There was a cascade of wonderfully clear, bright blue where a flax plant threw up a fountain of blossom. It was delightful. Why hadn't I fully appreciated it before?

I placed one hand over my left eye: the blossom vanished, and only when I approached close to the plant could I discern the flowers, now a greenish-grey shading into the foliage; I covered my right eye, and the wonderful blue returned. I looked up at the tall hawthorn hedge now in full bloom, its branches covered in dazzling, snow-white flowers. I covered the other eye. This time the blossom did not disappear; instead, it went a dingy yellow. Bluebells growing under a lilac tree were in full flower, but their flowers were almost as invisible as those of the flax when I covered my left eye. They could just be seen, darker than their foliage. It was with great pleasure that I looked at them with my other eye. Though they were much darker than the flax, their colour was a revelation.

Of course, I had just returned from the Wolverhampton Eye Infirmary, after a stay of four days during which a cataract had been removed from my left eye. It must have been dirty yellow to so affect the blue flowers. The very dark glasses I had worn until entering the garden had concealed the pleasures in store. There were no more blue flowers to look at, in fact nothing else in the garden beds was in bloom. I then remembered the hugely colourful auriculas in the greenhouse. Their blooms were over, save for those of 'Apple Cross', a gold-centred alpine auricula with deep-crimson, shaded petals. This was the one disappointment. The golden centres were dazzling, far too gaudy for the velvety red of the rest of the petals. Their centres dominated the plants, as bright as the glowing golden buttercups in the adjacent field. Still, I consoled myself, come the spring there would be the blue show-selfs such as 'Blue Jean', 'Stant's Blue', and 'Foreign Affair', as well as the alpines 'C. W. Needham', 'Frank Crossland', 'Mrs Hearn' and 'Roxburgh'.

I looked down and saw my socks. Horror! One was royal blue and the other black – I had worn the dark glasses when putting them on that morning. I went and looked at all the socks in the drawers upstairs. Grey, black, blue and brown were all mixed up together. Why had no one told me? (I am a widower and live alone.)

My reaction to the flax blooms was of course emotional, yet I had always considered myself to be unemotional. I recalled a discussion I had had with a psychology student some years before. He contended that reaction to all colours was emotional. If deep, bright colours (but not pastel shades) could be isolated from everything else – shape, past experience, associations, et cetera – all emotionally stable adults, whether in England, Africa or elsewhere, would (he claimed) prefer either blue or red, with green in third place. Children and the emotionally immature would prefer yellow or orange, while the emotionally disturbed would prefer purple, brown or black. Perhaps my delight in the blue flowers reflected my maturity. I was vain enough to hope so, though I have also found some purple auriculas, such as 'Lilac Domino', very pleasing. I have also been very glad to see glossy-black cock blackbirds in the garden. Can I be both mature *and* disturbed? That of course was a nonsense thought. The auriculas have form and elegance as well as colour, and the blackbird's activities and song are enough to override any colour reaction. Still, I have always wondered about those who profess great admiration for all-yellow or all-white herbaceous borders. I must banish such silly thoughts, concentrate on enjoying the flax, and perhaps read again what Gertrude Jekyll wrote about colour in the border.

For some months before my visit to the Eye Infirmary, I had been a little alarmed by those who, taking my age into account (approaching seventy-five), considered I ought to try to grow old gracefully. I reacted strongly to this idea, for any such an attempt would surely lead to the dreaded slippery slope that ends in vegetable senility. Surely it would be better to aim deliberately at a little irascibility, even at the risk of ending up a grumpy old curmudgeon? No sooner was this resolved upon than to hand came an excavation report that I strongly objected to. Carrying out my resolve, I sent the editor of an archaeological journal a broadside on the matter. The paper has been accepted, and though the editor insisted that some phrases be watered down, it is a step on the path I intend to follow. Or at least, I *had* intended, until I saw the flax and the bluebells. Perhaps after all it is too soon to degenerate into contentious bad temper. The new appearance of those blue flowers has made me feel years younger, and almost cheerful. My priority now must be to search the catalogues for more blue-flowering plants to add to my garden.

JIM GOULD, 1993

Where I live in a block of flats we share a patch of meadow which we are slowly – Rome wasn't built in a day – turning into a garden. We are turning it into a garden now that we have a fence and an entryphone system. Before this advance on the sunny uplands of civilization there was a period of chaos and old night when, open to the four winds and impromptu games of football (and worse), large dark dogs with faces that put one in mind of Orson Welles made free of the space and, to be frank, white honesty would not have stood a chance. Three out of the eight residents decided to do a little gardening, while the rest were torpidly indifferent. My place in the sun was a rectangle exactly the size of a double bed. Nor were the dimensions a matter of chance. During the period of chaos and old night someone had dumped a mattress on the green and then walked without taking up their bed. A day or two passed, then two larky youths thought it would be a merry jape to set the mattress on fire. It burned surprisingly well, too well; alarmed, the two boys tried to extinguish the flames in a manner which was a tribute to their imagination rather than a testimony to their sense of decorum. I caught them at it when going to my bedroom window to investigate the cause of the mephitic vapours which were tickling my nostrils. The Fire Brigade put out the flames, and when we had collected all the metal springs there was this large burnt patch. I daresay that in the entire history of gardening no other garden plat was selected by such a curious event. Well, I planted some roses and some peonies and other good things and excitedly checked every few minutes to see how things were going on. One of the other gardeners (I will not use her real name; let us call her La Malizia) advanced on me as I contemplated the patch, her face crinkled in a clumsy mask of sympathy, and said, 'Oh, John, I'm sorry to see that one of your roses has got suckers already; better get them off now, don't you think? Before they get a hold?' (La Malizia frequently tells anyone who will listen that she is a 'great' gardener.) 'Suckers?' I said, 'Surely not.' After all, the damned roses had only been in a matter of days. She loomed closer. 'There!' she said, indicating some red sprouts quite a foot away from the rose. It was an exquisite moment. I savoured it.

'No, Malizia, dear. Not suckers. A peony popping through.'

'Oh, a peony. *What* a relief!' she replied, in a thwarted tone of voice. Now there is a person who, if ever she crept into the Hardy Plant Society, would be frog-marched out middling smartish, I reckon. I feel better now I've told you that. Time to leaf through the Yellow Book, I fancy.

JOHN FRANCIS, 1991

Round the sky-blue lakes of Kashmir, the great Mogul Emperors once laid out their terraced gardens as a retreat from the wearying heat of India's dusty plains. Shalimar is the best-known name, but it is not romantically restored and maintained. 'We laid out gardens with order and symmetry, with proper parterres and borders in every corner and in every border, roses and narcissus in perfect arrangement . . .' For the echoes of a garden as Babur, the first Mogul Emperor, knew and planned one, there is more to be sensed in the nearby Nishat Bagh, laid out by a courtier of his successor Jahangir, probably in the 1620s.

Beneath its canopy of ageing plane trees, where a small roof-top iris grows wild on the terraced walls, I walked once in this garden's midday shade, looking out past its ranks of annual flowers to the serene middle distance of its planned background, the Dal Lake and its distant fort. This magical garden struck a curious note of discord, matching trees from the past Mogul centuries with cosmos daisies and violas sown from subsequent seed. Entering the sun, I met with a man who inquired, in well-phrased English, why his plants were of such interest to my eye. I had strayed from the terraces into the seedbeds of the chief gardener, employed at Nishat for the past forty years. We exchanged our views on jasmine and on the times when seedlings were best pricked out. He showed me the wild white roses which he grafted and I asked the names of his floating lotus. On a long white cloth, spread like a carpet beneath the chenar trees, he invited me to share a gardening lunch. I crossed legs at the head of this tablecloth of honour and sat eating curry with a scoop of the hand while his twenty-two junior gardeners watched with puzzlement on either side.

Down the length of the cloth, we swapped stories of annual flowers which my host then translated for his interested staff. Until 1947, he explained, the British ladies had assisted in the garden's planting. Here, then, lay the cause of that curious discord, for the style of post-war gardening in Surrey had been mixed with Nishat's natural prospect, once sensed by Jahangir's court. When independence took Kashmir by surprise, the British left in a hurry and took their seed-packets with them. Only the bedding plants of 1947 had survived the summer, and were now kept alive from year to year by the careful harvesting of their own seeds. By the Dal Lake I had caught a distant echo of the Englishwoman's garden. The clock, here, had stopped before F1 Hybrids, Glitter Petunias and Snapdragons with ruffled flowers. These losses, perhaps, were not so serious, but the news of other selections struck home like words of a long-lost paradise. Modern pansies, I called down the tablecloth, now come in blues and whites, while Zinnias bear lime-green flowers and Nasturtiums grow six inches high. The wonders of Suttons and Thompson & Morgan were translated for the younger gardeners' benefit. They murmured at 'pansies in separate colours' and applauded the mention of Love-in-a-mist in shades of pink and white. We arranged to send seeds, but feared the Kashmiri customs' control. Perhaps Nishat Bagh is alive now with night-scented stocks and selected strains of Dianthus. But I suspect these flowers may be blooming in a bureaucrat's urban backyard.

ROBIN LANE FOX, 1982

R ichard Andrews (1798–1850), son of a wheelwright, himself a coachbuilder of renown, friend of the deserving poor (how nice to be sure), prosperous and generous, owner of a manufactory in Above Bar, Southampton and five times – eat your heart out, Dick Whittington – Mayor of that county, now city, enjoys an enviable position in death, as he seems to have done in life. There he stands, a bit nibbled by time and car fumes. Surely it is the pierced state of the ozone layer, old coal fires and acid rain that have rendered so lacy those parts of his mayoral gown that were not intended to be lacy?

As I say, his position is enviable. To his left and to the right stretch avenues of fine horse chestnuts. Behind him, over his right shoulder, stands the poignant monument to the crew of the *Titanic*, realized in slippery-looking granite. Just to the right of the nape of his neck there live the sardonic-eyed, mocking élite of the aviary, much given to derisive bouts of laughter, struttings, twitterings, wolf-whistles; every last one of them glad to have won a place in the Royal Box, from which to observe the 'fairly' throbbing heart of Southampton, safe from cats and totally confident about the timing and the quality of the next meal.

But Richard Andrews' most desirable prospect is directly in front of those sightless eyes: the great Lime Walk stretching, if not as far as the eye can see, at least as far as a twenty-minute stroll takes. I see Mayor Andrews' statue every morning when I pass him on my way home. Home, or kipperwards, or boiled eggwards, even, in Babylonic mood, tomato and baconwards, after a night's work. Praising God and eyeing the hostas, you may say. That is, if I go by way of Brunswick Place.

Brunswick Place is the edging for the northern end of the parks. I say parks, for though they form a whole, it is to be hoped that tiresome roads will not punctuate them. I see Mayor Andrews; perhaps he sees me. It is more likely, for the little I know of him inclines me to suppose that he was a discriminating man, that he is regarding the vista of the Lime Walk. The late hurricane saw off several limes. I have noticed one adolescent replacement, planted to commemorate the life of Norman Dalrymple, cabinetmaker, born 29 November 1938, killed Clapham rail crash, 12 December 1988. How much better in this tragic circumstance to plant a tree than a stone. And a cabinetmaker surely would be pleased to know that more wood, so to speak, was on the way. I digress.

Parks are conducive to digressing.

The statue of Lord Palmerston, Prime Minister, Foreign Secretary, keen dispatcher of gun-boats, has a charming spot all its own, which he keeps an eye on. Such a mild eye, such a milky, innocent look! Clad in a dressing-gown with toga-ish overtones he seems to be in charge of a rockery, winding paths and pretty flowering shrubs. If all one has heard is half true, he would be indulgent towards any lascivious goings-on in this section.

I admit that as I walk home at about six in the morning five days a week I am sometimes seized with a mild and harmless sense of megalomania. 'All mine!' The Siren voices of the kipper and the boiled egg are silenced. I feel at one with Louis XIV, whom one can imagine, roused by his faithful body-servant Bontemps, looking out after the curtains had been thrown back and seeing vistas. And not just vistas. With his inward eye he would also have seen another lovely day of sheer Kingship: snubbing a pushy courtier here, graciously receiving a humble petition there, having a spousy chit-chat with his sanctimonious old humbug of a second wife before invading somewhere; but before doing any of these mundane things, seeing the amazing park that he had created with the help of his wily old Le Nôtre. The advantages of being Louis XIV seem very real when, sharing it with the dead eyes of Richard Andrews, you see the Lime Walk. Sets you up for a good day's sleep. And, of course, you got up twice: once for yourself, then officially, as King. 'Dawn broke in the Royal Bed-chamber at . . . fill in the space yourself, your Majesty.' Then at a signal (this must have been a bit trying) a Royal flourish of fifes and drums alerted the waiting world – well, Versailles anyway – to the tell-tale tremor of the Royal eyelids. 'Yes, Sire, ghastly din but just swallow this aspirin, and the roses are looking lovely.'

At sixish the parks are all but deserted. All but. Once from out of a thicket of rhododendrons I was surprised by a vivid-voiced, red-faced old party bidding me 'Good morning' and asking me what the time was. He was still cheerful and the source of his cheerfulness, bottled Somerset orchards, lay strewn about him.

Of course, the fancy that you are the Emperor of China, le Roi Soleil or Lord Tom Noddy is forever being ambushed. For you are not really alone. Tulips like hock glasses filled with the wrong wine peer tremulously at you, borne on a tray of freakish mist which would be dispersed by the errant flight of a butterfly. You know the sort of local mist I mean; it clings over streams, it heralds pop groups on tele-vision; and it is regularly waded through by Vincent Price in a battery of old horror movies. Stand still, heedless of tomatoes and bacon: this is not a repeatable experience. It is not for storing on even the most sophisticated video machine, and banality lies in wait. Plump, plump, plump. Two stout ladies in track suits, keenly anxious to shed those ugly pounds, thud past the aviary. The peacocks, the cockatiels, and me, distinctly hear one say (for even exercise must not dethrone gossip, thank goodness) 'I don't see how she can be a bridesmaid, not with those tattoo'd arms.'

Having no garden of my own, unless you count my window boxes, I number the parks of Southampton, whatever the Parks Department fondly imagine, among *my* most cherished possessions. They are a noted glory, conceded by even the most biliously hostile dweller in neighbouring Portsmouth to be not half bad. Not that they are per-fect. I appreciate that parks, being large-scale, necessarily paint in bold colours. Even so, I question faintly the wisdom of egg-yolk yellow wallflowers topped off with puce tulips. Again, fiendish experiments by mad horticulturists who produce regrettable pelargoniums should be resisted, not encouraged. It's odd the way Parks people seem to long for a vivid tropical look to northern European gardens. I've noticed bits of Cheltenham, of all places, attempting to pass themselves off as Brazil. Better to have a space of uncut grass and a few primroses; and what I often ask is wrong with foxgloves?

I like rhododendrons, but I'm against those blue azaleas that seem to be plugged into the national grid.

> If thou wouldst view fair Melrose aright,
> Go visit it by the pale moonlight.

If thou wouldst view Southampton's parks, moonlight might reveal some growths of a monstrous aspect which are best avoided. No. The best time to view parks is a little after six in the morning in early summer, more particularly if you work at night and have no garden of your own. Then it's paradise.

JOHN FRANCIS, 1989

It's spring in the city, April in New York, and the crack gardeners – the ones who draw their horticultural inspiration from the dirt-cheap, lethal cousin of cocaine – are back working the soil and the street. Plant muggings on this block already include a purple primrose by the roots and three hyacinths by the bulbs. Crack gardeners finger plants for their resale value. A vial of crack goes for $2.50 or so in this borough: an azalea is two vials of crack, a rose three or four.

According to a June 1988 *Time* cover story, quite a lot of Americans are into gardening these days, and, according to a *Time* cover just a month earlier, quite a lot of Americans are also into crack. *Time*, however, failed to make the crack/gardening connection. I will, because my Brooklyn street garden is situated smack on the frontier of this new urban American gardening trend.

Crack gardeners care about ornamental plants only in so much as the plants can be cashed for crack, and so they prefer upscale shrubs and perennials. In inner-city neighborhoods such as mine, crack gardening has given rise to counter-crack gardening. For this reason, urban gardening must be reckoned among the most advanced of our indigenous gardening schools. (Nevertheless, counter-crack gardening has yet to be acknowledged by any experts on urban gardening.) Counter-crack gardening requires a radical change of style. My street gardening has become combative, thorny and downright offensive – I am plotting a plot of thoroughly vicious plants.

A year ago the first crack gardeners hit my neighborhood, tearing apart the charming street garden of a floral designer. Over the course of a week, working weeknights, they stole a weighty terracotta planter filled with herbs and flowers; a flatsworth of lettuce; a white azalea; and roses, among them a blooming, pink 'Betty Prior'. They hacked at the limbs of a Rose of Sharon and left the branches dangling by a thin bark skin. They uprooted a border of twenty-four pink wax begonias. They cut the chains securing some large, top-of-stoop planters and disappeared them.

Across the street and around the corner the MIAs included: three azaleas, one *Pieris japonica*, euonymus bushes, roses and three 'Enchantment' lilies. Crack gardeners are not mere blossom snatchers – they appreciate the higher resale value of the whole plant, and always steal the roots.

At the height of the crack gardening spree, officer Bert Marrero of Brooklyn's 77th Precinct told me: 'I just locked up a guy for stealing some shrubbery. He was a crack addict, trying to sell the shrubbery to the merchants on Flatbush Avenue . . . Crack is an inexpensive drug. With a few bucks they can make a buy. So they're rippin' little things off . . . Don't be surprised if they take your plants and we don't get the perps.'

I did not much like 'Cary Grant', the rose, when we met at the 1988 New York Flower Show – an orange-juice-colored, fragrantless modern. But crack came into the garden and my neighbor lost a rose. I suggested 'Cary' – a hyper-thorny hybrid tea, a brass knucks of a rose. 'Cary Grant' lasted two weeks in Brooklyn.

It is possible that the nursery business will do as well by the crack epidemic, considering the garden fallout, as has, inadvertently, the baking soda biz. For instance, a gardening friend, fed up with crack plant predators, last spring ordered 100 'Simplicity' roses from Jackson & Perkins, which guaranteed the protection of '100 feet of Simplicity Living Fence'. A year later, thirty-seven 'Simplicity' roses remain. In the interim, various of the crackheads who had done the stealing approached the gardener with excellent buyback deals.

This woman had begun the garden in 1986 to give street kids something to do. By 1988 she knew that several of her little gardeners also were working as crack runners. She lost some kindergarterners when she decreed: 'Either garden or run drugs – you can't do both!'

Last year during fall 'bulbing', instead of squirrels digging up her bulbs, crack-heads dug them – and tried to sell them back to her. And one autumn day the crack gardeners made off with a hand truck, seven spades, seventeen trowels, hand cultivators and shovels. 'I am not ready to put my life on the line for gardening tools,' my friend informed me.

Crack is a pestilence for which there is no pesticide. Urban counter-crack gardeners begin to envy suburban gardeners their kinder, gentler pests – aphids, grubs, bugs, slugs. Nowhere ... do I find Gertrude Jekyll, Vita Sackville-West, William Robinson, E. A. Bowles or Reginald Farrer complaining of anything more noxious than slugs and snails. Nor do any of the thirteen authors in the Pests and Poisons chapter of *The American Gardener* deal with anything more pesky than children and apple tree borers.

Some counter-crack gardeners dream about booby-trapping the petunias. A neighbor recalls the tactical value of upfront blackberries in his East Harlem youth, 'Blackberries! That's the thing! Nobody'd touch those vicious bushes!' Enraged urban green thumbs suggest glass-shard and razor-wire mulches. Appraising your plants in terms of crack, you realize that not only must you 'plant down', you must plant mean. (This garden won't be entered in the National Peace Garden competition in Washington, D. C.)

Considering there were 697 murders in Brooklyn last year, and 1,896 citywide, crack gardening is a mere blip on the city's crime screen. It is really up to counter-crack gardeners to help themselves, mindful of garden guru Christopher Lloyd's dictum that we need to be aware 'of the uses to which armed plants can be put as offensive weapons'. This spring my defense-procurement plant contracts have been let, solely on the basis of each plant's reputation for viciousness, and the heck with competitive bidding. Even now the following defensive-garden weapons systems are in the pipeline.

For the rear and side defensive perimeter: Hardy Orange (*Poncirus trifoliata*) – a wickedly thorny shrub that, according to *Taylor's Encyclopedia of Gardening*, 'forms impenetrable, defensive hedges'; Hawthorn (*Crataegus*); Blackberry (*Rubus*); the Sweetbriar rose (*Rosa eglanteria*) – 'the stems bristly, and with strongly hooked prickles', *Taylor's* says; Father Hugo's rose (*Rosa hugonis*) – quoth *Taylor's* 'beset with flattened, straight prickles and bristles'; the equally prickly Scotch rose (*Rosa spinosissima*).

For the front defensive perimeter: Barberry (*Berberis*); Prickly Pear cactus (*Opuntia compressa*); and, in extremis, Poison Ivy (*Rhus toxicodendron*). For the defensive interior: Scotch thistle (*Onopordum acanthium*); a Sea Holly (*Eryngium yuccifolium*, a.k.a. rattlesnake master); Burdock (*Arctium lappa*); Stinging Nettle (*Urtica dioica*) and Deadly Nightshade (variously *Solanum nigrum* or *S. dulcamara*). These plants are plenty nasty enough for the offensive gardening necessary to the counter-crack attack, even if the garden will paint a pretty prickly picture.

PATTI HAGAN, 1989

EPILOGUE

But the Quincunx of Heaven¹ runs low, and 'tis time to close the five ports of knowledge; We are unwilling to spin out our awaking thoughts into the phantasmes of sleep, which often continueth praecogitations; making Cables of Cobwebbes and Wildernesses of handsome Groves. Beside *Hippocrates* hath spoke so little and the Oneirocriticall Masters, have left such frigid Interpretations from plants, that there is little encouragement to dream of Paradise itself. Nor will the sweetest delight of Gardens afford much comfort in sleep; wherein the dulnesse of that sense shakes hands with delectable odours; and though in the Bed of *Cleopatra*,² can hardly with any delight raise up the ghost of a Rose.

Night which Pagan Theology could make the daughter of *Chaos*, affords no advantage to the description of order: Although no lower then that Masse can we derive its Genealogy. All things began in order, so shall they end, and so shall they begin again; according to the ordainer of order and mystical Mathematicks of the City of Heaven.

Though *Somnus* in *Homer* be sent to rowse up *Agamemnon*, I finde no such effects in these drowsy approaches of sleep. To keep our eyes open longer were but to act our *Antipodes*. The Huntsmen are up in *America*, and they are already past their first sleep in *Persia*. But who can be drowsie at that howr which freed us from everlasting sleep? or have slumbring thoughts at that time, when sleep itself must end, and as some conjecture all shall awake again?

SIR THOMAS BROWNE, 1658

1 *Hyades* near the Horizon about midnight, at that time [footnotes as in original].
2 Strewed with roses.

BIBLIOGRAPHY

Addison, Joseph, 1710, *The Tatler*, London

—— 1712, *The Spectator*, London

Akeroyd, John, 1991, 'In Praise of Some Autumn-flowering Bulbs', *Hortus* (No. 18)

Allsopp, Jane, 1994, 'English, Perennial, Rustic, and Alone', *Hortus* (No. 30)

Armitage, Ethel, 1936, *A Country Garden*, London

Austen, Jane, 1816, *Emma*, London

Austin, Alfred, 1894, *The Garden That I Love*, London

Bacon, Sir Francis, 1625, 'Of Gardens', *Essays*, London

Baring-Gould, S., 1890, *Old Country Life*, London

Bates, H. E., 1971, *A Love of Flowers*, London

Beaton, Mr, 1841, in *Ladies' Magazine of Gardening*, 1843

Beckford, William, 1782–3 (quoted in *The Grand Tour*, Roger Hudson, London, 1993)

Binyon, Laurence, 1944, *The Burning of the Leaves and Other Poems*, London

Blaikie, Thomas, 1786 (quoted in *The Pleasures of Diaries*, Ronald Blythe, London, 1989)

Blake, Stephen, 1664, *The Compleat Gardeners Practice*, London

Blake, William, 1794, 'The Sick Rose', *Songs of Experience*, London

Blomfield, Sir Reginald, 1892, *The Formal Garden in England*, London

Blunt, Lady Anne, 1881, *A Pilgrimage to Nejd*, London

Blunt, Wilfrid, 1963, *Of Flowers and a Village*, London

Bowles, Edward Augustus, 1914, *My Garden in Spring*, London

——1914, *My Garden in Summer*, London

——1915, *My Garden in Autumn and Winter*, London

Boyle, Mrs E. V., 1884, *Days and Hours in a Garden*, London

Brontë, Charlotte, 1853, *Villette*, London

Brough, John (trans.), 1968, *Poems from the Sanskrit*, London

Browne, Sir Thomas, 1658, *The Garden of Cyrus*, London

Burnet, Gilbert (Bishop of Salisbury), 1686, *Some Letters, containing an Account of what seemed most remarkable in Switzerland, Italy, etc.* (quoted in *Italian Villas and Their Gardens*, Edith Wharton, New York, 1904)

Čapek, Karel, 1931 (English trans.), *The Gardener's Year*, London

Chardin, Sir John, 1686, *Travels in Persia 1673–1677*, London

Chatto, Beth, 1988, *Beth Chatto's Garden Notebook*, London

Christopher, Thomas, 1989, *In Search of Lost Roses*, New York

Clare, John, 1808–9, 'The Wish', in *The Oxford Book of Garden Verse*, Oxford, 1993

Cobbett, William, 1829, *The English Gardener*, London

Cowper, William, 1785, 'The Task', in *The Oxford Book of Garden Verse*, Oxford, 1993

Cran, Marion, 1929 (2nd edn), *The Joy of the Ground*, London

Dean, Richard, *c.* 1890, in *Cassell's Popular Gardening*, London

Denham, Humphrey John, 1940, *The Skeptical Gardener*, London

Dickens, Charles, 1852–3, *Bleak House*, London

Douglas, David, 1914, *Journal*, London

Downing, Andrew Jackson, 1841, *A Treaty on the Theory and Practice of Landscape Gardening*, New York

Dutton, G. F., 1995, *Harvesting the Edge*, London

Earle, Alice Morse, 1901, *Old Time Gardens*, New York

Earle, Mrs C. W., 1897, *Pot-Pourri from a Surrey Garden*, London

Eck, Joe, and Winterrowd, Wayne, 1995, *A Year at North Hill: Four Seasons in a Vermont Garden*, New York

Eden, Frederick, 1903, *A Garden in Venice*, London

Eliot, George, 1858, *Scenes of Clerical Life*, London

Ellacombe, Canon Henry, 1895, *In a Gloucestershire Garden*, London

Evelyn, John, 1664, *Sylva*, London

—— 1699, *Acetaria, A Discourse of Sallets*, London

——1818–19, post., *Diary*, London

—— 1932, post., *Directions for the Gardener at Sayes-Court*, London

Fairchild, David, 1939, *The World Was My Garden: Travels of a Plant Explorer*, New York

Fairchild, Thomas, 1722, *The City Gardener*, London

'Falconer, Christopher', 1969 (pseudonym, quoted in *Akenfield*, Ronald Blythe, London)

Farrer, Reginald, 1921, *The Rainbow Bridge*, London

Fearon, Ethelind, 1952, *The Reluctant Gardener*, London

Fish, Margery, 1956, *We Made a Garden*, London

——1961, *Cottage Garden Flowers*, London

——1965, *A Flower for Every Day*, London

Fletcher, H. L. V., 1948, *Purest Pleasure*, London

Forbes, Alexander, 1820, *Short Hints on Ornamental Gardening*, London

Fortune, Robert, 1847, *Three Years' Wanderings in China*, London

Fox, Robin Lane, 1982, *Better Gardening*, Beckley, Oxfordshire

Francis, John, 1989, 'Paradise in the Parks', *Hortus* (No. 11)

——1991, 'Playing One's Part', *Hortus* (No. 18)

Gentil, François, and Liger, Louis, 1706, *The Retir'd Gard'ner* (trans. by George London and Henry Wise), London

Gerard, John, 1597, *Herball*, London

Gilbert, Samuel, 1682, *The Florists Vade-Mecum*, 3rd edn, London, 1702

Gissing, George, 1903, *The Private Papers of Henry Ryecroft*, London

Goodall, Nancy-Mary, 1987, 'Gardens in Fiction: *The Aspern Papers*', *Hortus* (No. 2)

Gould, Jim, 1993, 'Blue Pleasure', *Hortus* (No. 27)

Grant, Anne, 1808, *Memories of an American Lady*, London

Grigson, Geoffrey, 1952, *Gardenage: or The Plants of Ninhursaga*, London

Hagan, Patti, 1989, 'Countering the Crack Attack', the *Wall Street Journal*

Haggard, H. Rider, 1905, *A Gardener's Year*, London

Hancock, Thomas, 1737 (quoted in *A History of Horticulture in America*, U. P. Hedrick, Oregon, 1950)

Hanmer, Sir Thomas, 1659, *The Garden Book of Sir Thomas Hanmer* (first pub. 1933, London)

Harding, Alice, 1923, *Peonies in the Little Garden*, New York

Hardy, Thomas, 1880, *The Trumpet-Major*, London

——1891, *Tess of the d'Urbervilles*, London

Herbert, Sir Thomas, 1678 (quoted in *The Trial of Charles I*, ed. Roger Rockyer, London, 1959)

Hibberd, Shirley, 1856, *Rustic Adornments for Homes of Good Taste*, London

Hill, Jason, 1932, *The Curious Gardener*, London

——1940, *The Contemplative Gardener*, London

Hill, Thomas, 1577, *The Gardener's Labyrinth*, ed. Richard Mabey, Oxford, 1987

Hole, S. Reynolds, 1885, *A Book about Roses*, London

——1899, *Our Gardens*, London

——1909, *A Book about the Garden*, London

Honey, William Bowyer, 1939, *Gardening Heresies and Devotions*, London

Jarvis, Major C. S., 1951, *Gardener's Medley*, London

Jefferson, Thomas, quoted in *The Garden and Farm Books of Thomas Jefferson*, ed. Robert C. Baron, Golden, Colorado, 1987

Jekyll, Gertrude, 1899, *Wood and Garden*, London

——1900, *Home and Garden*, London

Johnes, Thomas, 1804, *A Land of Pure Delight: Selections from the Letters of Thomas Johnes of Hafod, Cardiganshire*, ed. Richard J. Moore-Colyer, London, 1922

Johnson, Arthur Tysilio, 1950, *The Mill Garden*, London

Jones, H. Clayton, 1951, 'Gardening Terms of the Welsh Marches', *My Garden* (No. 209)

Jones, Henry, 1749, 'On a Fine Crop of Peas being Spoiled by a Storm', in *The Oxford Book of Garden Verse*, Oxford, 1993

Kellaway, Deborah, 1993, 'Pugs, Peacocks and Pekingese: The Garden at Garsington Manor', *Hortus* (No. 25)

Kelly, John, 1992, 'Gardening Myths and Commandments', *Hortus* (No. 21)

Kipling, Rudyard, 1911, 'The Glory of the Garden', *Rudyard Kipling's Verse: Definitive Edition*, London

Knox, Vicesimus, *c.*1778, *Essays, Moral and Literary* (quoted in *The Genius of the Place: The English Landscape Garden 1620–1820*, John Dixon Hunt and Peter Willis, London, 1975)

Krutch, Joseph Wood, 1954, *The Voice of the Desert*, New York

Lacey, Stephen, 1989, 'Stephen Lacey's Snippets', *Hortus* (No. 10)

Langley, Batty, 1728, *New Principles of Gardening*, London

Laurence, John, 1726, *A New System of Agriculture, Being a Complete Body of Husbandry and Gardening*, London

Lawrence, Elizabeth, 1965, '"Cameelias" and Such' (reprinted in *Through the Garden Gate*, Chapel Hill, North Carolina, 1990)

Lawson, William, 1618, *A New Orchard and Garden*, London

Leighton, Clare, 1935, *Four Hedges*, London

Lodwick, John, 1960, *The Asparagus Trench*, London

Loudon, Jane, 1841, *Practical Instructions in Gardening for Ladies*, London

Loudon, John Claudius, 1829–33 (quoted in *In Search of English Gardens: The Travels of John Claudius Loudon and His Wife Jane*, ed. Priscilla Boniface, London, 1990)

Lytton, Bulwer, 1837, *Ernest Maltravers*, London

McEacharn, Neil, 1954, *The Villa Taranto: A Scotsman's Garden in Italy*, London

MacLeod, Dawn, 1994, 'Looking Back', *Hortus* (No. 29)

Markham, Gervase, 1638 edn, *A Way to Get Wealth*, London

Martineau, Alice, 1923, *The Herbaceous Garden*, London

Marvell, Andrew, *c.*1651, 'The Garden' (first published 1681, London)

Meager, Leonard, 1688, *The English Gardner*, London

Milne, A. A., 1920, *If I May*, London

Mitchell, Henry, 1981, *The Essential Earthman*, Indiana

Morris, William, 1882, *Art & Social Reform*, London

Nichols, Beverley, 1932, *Down the Garden Path*, London

Nicolson, Harold, 1966–8, *Harold Nicolson's Diaries*, 3 vols., ed. Nigel Nicolson, London

Oliphant, Mrs, 1866, *Miss Marjoribanks* (quoted in *The Garden in Victorian Literature*, M. Waters, Aldershot, 1988)

Origo, Iris, 1970, *Images and Shadows*, London

Osler, Mirabel, 1990, 'A Word about Boxes', *Hortus* (No. 16)

Page, Russell, 1962, *The Education of a Gardener*, London

Palgrave, Francis Turner, 1871, 'Eutopia', in *The Oxford Book of Garden Verse*, Oxford, 1993

Parker, Peter, 1994, 'Peter Parker Ponders . . .', *Hortus* (No. 31)

Parkinson, John, 1629, *Paradisi in Sole Paradisus Terrestris*, London

Paxton, Sir Joseph, 1826 (quoted in *A Century of Gardeners*, Betty Massingham, London, 1982)

Pemberton, Rev. Joseph, 1908, *Roses: Their History, Development and Cultivation*, London

Percy, Thomas (Bishop of Dromore), 1760 (quoted in *The Percy Letters: The Correspondence of Thomas Percy and William Shenstone*, ed. C. Brooks, New Haven, Connecticut, 1977)

Phillpotts, Eden, 1906, *My Garden*, London

Pitter, Ruth, 1941, 'The Diehards', in *The Oxford Book of Garden Verse*, Oxford, 1993

Pope, Alexander, 1720, letter to Robert Digby (quoted in *Alexander Pope: A Life*, Maynard Mack, New Haven and London, 1985)

——1725, letter to Edward Blount (quoted in *ibid.*)

Powell, Joseph, and Gribble, Francis, 1919, *The History of Ruhleben: A Record of British Organisation in a Prison Camp in Germany*, London

Ravenel, Harriott Horry, 1906, *Charleston: The Place and the People*, New York

Rea, John, 1676, *Flora, Ceres and Pomona*, London

Repton, Humphry, 1795, *Sketches and Hints on Landscape Gardening*, London

——1840, *The Landscape Gardening and Landscape Architecture of the Late Humphry Repton Esq* (ed. J. Loudon)

Ribbesford, Timothy, 1993, 'Folly Aforethought', *Hortus* (No. 23)

Robinson, Liz, 1991, 'Liz Robinson's Snippets', *Hortus* (No. 18)

Robinson, William, 1870 (5th edn, 1895), *The Wild Garden*, London

Rohde, Eleanour Sinclair, 1936, *Herbs and Herb Gardening*, London

Ruskin, John, 1875–86, *Proserpina*, London

Sackville-West, Vita (Victoria), 1937, *Some Flowers*, London

—— 1954, letter quoted in *Harold Nicolson's Diaries*, op. cit.

Scott, Frank, 1870, *The Art of Beautifying Suburban Home Grounds of Small Extent*, New York

Sedding, John Dando, 1890, *Garden-craft Old and New*, London

Shakespeare, William, *c.*1595, *Richard II*, London

Shebbeare, John, 1755, *Letters on the English Nation*, London

Sister of the Shaker Community, 1906 (quoted in *A History of Horti-culture in America*, U. P. Hedrick, Oregon, 1950)

Sitwell, Sir George, 1909, *An Essay on the Making of Gardens*, London

Sitwell, Sacheverell, 1948, *Old Fashioned Flowers*, London

Stevenson, Robert Louis, 1904, *Memories and Portraits*, London

——1905, post., *Essays of Travel*, London

Swinscow, Douglas, 1992, *The Mystic Garden*, Tiverton

Switzer, Stephen, 1731, *A Compendious Method for the Raising of Brocoli, Spanish Cardoon, Celeriac, Finochi, and Other Foreign-Vegetables*, London

Tanner, Heather and Robin, 1939, *Wiltshire Village*, London

Taylor, George, 1948, *The Little Garden*, London

Temple, Sir William, 1692, *Upon the Gardens of Epicurus, or of Gardening, in the Year 1685*, London

Tennyson, Alfred, 1847, 'The Flower', from *The Princess*, London

Thaxter, Celia, 1894, *An Island Garden*, Boston, MA

Thelwell, Norman, 1978, *A Plank Bridge by a Pool*, London

Thomas, Graham Stuart, 1983, *Three Gardens*, London

Thomson, William, *c.* 1890, in *Cassell's Popular Gardening*, London

Toms, Derek, 1995, 'Godless', *Hortus* (No. 36)

Trollope, Frances, 1832, *Domestic Manners of the Americans*, London

Tyas, Robert, 1842, *The Sentiment of Flowers*, London

Verey, Rosemary, 1988, *The Garden in Winter*, London

Waley, Arthur (trans.), 1946, 'Planting Flowers on the Eastern Embankment', *Chinese Poems*, London

Walpole, Horace, *c.* 1750–60, *The History of the Modern Taste in Gardening*, London

Ward, Frank Kingdon, 1935, *The Romance of Gardening*, London

Warner, Charles Dudley, 1876, *My Summer in a Garden*, London

Waterfield, Margaret, 1907, *Flower Grouping in English, Scotch and Irish Gardens*, London

White, Katharine S., 1979, *Onward and Upward in the Garden*, New York (originally published in *The New Yorker*, 24 September, 1960)

Whittle, Tyler, 1965, *Some Ancient Gentlemen*, London

——1969, *Common or Garden*, London

Wilder, Louise Beebe, 1918, *Colour in My Garden*, New York

Wildsmith, William, *c.* 1890, in *Cassell's Popular Gardening*, London

Young, Andrew, 1985, 'Cornish Flower-Farm', *The Poetical Works*, London

INDEX OF WRITERS

GENERAL INDEX

READ MORE IN PENGUIN

In every corner of the world, on every subject under the sun, Penguin represents quality and variety – the very best in publishing today.

For complete information about books available from Penguin – including Puffins, Penguin Classics and Arkana – and how to order them, write to us at the appropriate address below. Please note that for copyright reasons the selection of books varies from country to country.

In the United Kingdom: Please write to *Dept. EP, Penguin Books Ltd, Bath Road, Harmondsworth, West Drayton, Middlesex UB7 0DA*

In the United States: Please write to *Consumer Sales, Penguin Putnam Inc., P.O. Box 999, Dept. 17109, Bergenfield, New Jersey 07621-0120.* VISA and MasterCard holders call 1-800-253-6476 to order Penguin titles

In Canada: Please write to *Penguin Books Canada Ltd, 10 Alcorn Avenue, Suite 300, Toronto, Ontario M4V 3B2*

In Australia: Please write to *Penguin Books Australia Ltd, P.O. Box 257, Ringwood, Victoria 3134*

In New Zealand: Please write to *Penguin Books (NZ) Ltd, Private Bag 102902, North Shore Mail Centre, Auckland 10*

In India: Please write to *Penguin Books India Pvt Ltd, 210 Chiranjiv Tower, 43 Nehru Place, New Delhi 110 019*

In the Netherlands: Please write to *Penguin Books Netherlands bv, Postbus 3507, NL-1001 AH Amsterdam*

In Germany: Please write to *Penguin Books Deutschland GmbH, Metzlerstrasse 26, 60594 Frankfurt am Main*

In Spain: Please write to *Penguin Books S. A., Bravo Murillo 19, 1° B, 28015 Madrid*

In Italy: Please write to *Penguin Italia s.r.l., Via Benedetto Croce 2, 20094 Corsico, Milano*

In France: Please write to *Penguin France, Le Carré Wilson, 62 rue Benjamin Baillaud, 31500 Toulouse*

In Japan: Please write to *Penguin Books Japan Ltd, Kaneko Building, 2-3-25 Koraku, Bunkyo-Ku, Tokyo 112*

In South Africa: Please write to *Penguin Books South Africa (Pty) Ltd, Private Bag X14, Parkview, 2122 Johannesburg*

GARDENING

The Well-Tempered Garden   Christopher Lloyd

A thoroughly revised and updated edition of the great gardening classic.

'By far the best-informed, liveliest and most worthwhile gardener–writer of our time ... There is no reasonable excuse for any gardener failing to possess Christopher Lloyd's books' – *Interiors*

The Confident Gardener   Brian Davis

Becoming a skilled gardener is a continual learning process and even the most experienced can always build on what they already know. In this highly informative book, Brian Davis, who has over thirty years of gardening experience, offers sound information and advice on how to keep your garden in top form throughout the year.

Organic Gardening   Lawrence D. Hills

The classic manual on growing fruit and vegetables without using artificial or harmful fertilizers. 'Enormous value ... enthusiastic writing and off-beat tips' – *Daily Mail*

GARDENING

The Adventurous Gardener   Christopher Lloyd

Prejudiced, delightful and always stimulating, Christopher Lloyd's book is essential reading for everyone who loves gardening. 'Get it and enjoy it' – *Financial Times*

Gardens of a Golden Afternoon   Jane Brown

'A Lutyens house with a Jekyll garden' was an Edwardian catch-phrase denoting excellence, something fabulous in both scale and detail. Together they created over 100 gardens, and in this magnificent book Jane Brown tells the story of their unusual and abundantly creative partnership.

A History of British Gardening   Miles Hadfield

From the Tudor knot gardens, through the formal realities of Jacobean and Georgian landscaping and on to the Gothic fantasies of wealthy Victorian landowners, Miles Hadfield brings the British gardens of the past vividly alive. 'An extraordinarily rich harvest of valuable and entertaining information . . . it is hard to see that it can ever be superseded' – *Journal of the Royal Horticultural Society*

Plants from the Past   David Stuart and James Sutherland

As soon as it is planted, even the most modern garden can be full of history, whether overflowing with flowers domesticated by the early civilizations of Mesopotamia or with plants collected in the Himalayas for Victorian millionaires. 'A thoroughly engaging style that sometimes allows bracingly sharp claws to emerge from velvet paws' – *World of Interiors*